Upstream Engineering and Flow Assurance Forum 2020

Held at the 2020 AIChE Annual Meeting

Online
16 - 20 November 2020

ISBN: 978-1-7138-2318-6

Printed from e-media with permission by:

Curran Associates, Inc.
57 Morehouse Lane
Red Hook, NY 12571

Printed with permission by Curran Associates, Inc. (2021)

For permission requests, please contact AIChE
at the address below.

AIChE
120 Wall Street, FL 23
New York, NY 10005-4020

Phone: (800) 242-4363
Fax: (203) 775-5177

www.aiche.org

Additional copies of this publication are available from:

Curran Associates, Inc.
57 Morehouse Lane
Red Hook, NY 12571 USA
Phone: 845-758-0400
Fax: 845-758-2633
Email: curran@proceedings.com
Web: www.proceedings.com

TABLE OF CONTENTS

608010 Acceleration of CFD Based Corrosion Models Using Machine Learning Algorithms

Abstract:

A computational fluid dynamics (CFD) model combined with electrochemistry has been developed to prediction CO_2 corrosion in pipelines [1]. The CFD model can explicitly account for corrosion chemistry with parameters of temperature, CO_2 concentration, pipe geometry and local fluid velocity. This allows a full parametric simulation of the maximum corrosion rate on steel pipes with arbitrary geometries. For general applications to work, commonly used pipe shapes with bends can greatly benefit the industry that allows real-time results for CO_2 corrosion predictions. However, the idea can be also extended to other geometrical shapes for various equipment and processing units prone to corrosion. Further extension to O_2 and H_2S corrosion, and multiphase flows can be highly beneficial as well. In this work, we introduce a real-time machine-learning (ML) model that replicates the CFD predicted maximum corrosion rate in pipes with various bend angles, pipe inner diameter and bend radius within a range of flow conditions. A total of 7 variables, including pipe ID, flow inlet velocity, pH value, CO_2 partial pressure, pipe bend angle, pipe bend radius and temperature, are parametrized to generate the maximum corrosion rate using CFD. A total of 83,160 simulation cases were performed using 7 ANSYS Fluent licenses with 48 CPU cores each, and 7 weeks of computational time.

A light gradient boosting (LGBM) ML algorithm [2] has been applied to model the CFD data. The optimized model gave a resulting R^2 value of 0.992 and a value of 0.975 for another new set of CFD data, indicating high efficiency of the model. The forward inferencing of the LGBM model only takes a few milliseconds on a laptop computer in comparison to the 10 minutes computing time of the CFD simulation using 48 CPU cores. The completed model was implemented in an interactive web tool where users can obtain results with graphics instantaneously. One interesting observation is that the web results can be more accurate than CFD simulations because the errors resulted from incorrect problem setup or numerical divergence can be avoided since the ML model always provides a locally filtered outcome that smoothes out potential erroneous results.

The value of this work lies in the wide range of applicability for general piping corrosion with CO_2. The proposed method can be extended to other corrosion mechanisms as well, e.g., O_2 and H_2S. However, corrosion is a complex phenomenon and results can vary significantly due to metal types, surface deposits and corrosion product films. Nevertheless, this method can be integrated with other more complex corrosion mechanisms and aid the scale-up or down of experiments or plant observations that cannot be easily obtained otherwise. Some examples will be provided and discussed, and other possible extensions will be addressed.

References:

[1] Kuochen Tsai, "Corrosion modeling using electrochemistry and computational fluid dynamics", 2017 AIChE Annual Meeting, Oct 31, Minneapolis, MN, USA, Paper No. 355b.

[2] Cheng Chen, Qingmei Zhang. QinMacBinYu, "LightGBM-PPI: Predicting protein-protein interactions through LightGBM with multi-information fusion", *Chemometrics Intell. Lab. Syst.*, vol. 191, pp. 54-64,2019.

609476 Machine Learning Approach to Accurately Model Corrosion Rates As a Function of Time

Abstract:

Internal and external corrosion in the oil and gas industry is a major concern. A widely-used method to lower the internal corrosion of oil pipelines is injecting corrosion inhibitors (CI) into the oil stream. Monitoring the corrosion rate as a function of time in the presence and absence of corrosion inhibitors is imperative to ensure that failures are timely detected. In the absence of a robust mathematical model to predict the corrosion rate as a function of time, frequent measurements of corrosion rates are performed, which are both expensive and time-consuming. In this study, we have employed different machine learning (ML) approaches to model the corrosion rate as a function of time for any different CI concentration and dose sequences based on the experimental data. We show that a Random Forest (RF) based ML model is accurately able to model the corrosion rate as a function of time. In this presentation, we will discuss the methodology of developing the ML model and the results obtained.

610546 Real Time Monitoring Technique for Detection of Microbial Corrosion in Oil and Gas System

Abstract:

As natural gas production in the United States continues to increase, the abundance of transmission pipelines will increase similarly to accommodate the productivity. These pipelines are susceptible to internal MIC, which is exceedingly difficult to detect. First, MIC is by nature a localized process, and localizing damage is nearly impossible with "bulk" pipeline interrogation techniques. Second, highly specialized techniques are required to reliably diagnose MIC. Third, detection of internal MIC requires removal of samples from the pipeline for analysis and therefore, downtime. Fourth, evidence of MIC may not be evident until the system is "infected" beyond recovery. We have developed an electrochemical approach that involves zero-resistance ammetry (ZRA) measurements along a pipeline to detect MIC. The ZRA approach could be deployed along entire pipelines and would sensitively detect a broad range of in situ microbiological processes.

To develop ZRA as an MIC monitoring tool, we have carried out a series of split chamber ZRA (SC-ZRA) incubations that entail deployment of two steel working electrodes (WE1 and WE2) in chambers separated by a semipermeable membrane. Subsequently, one chamber is inoculated with an individual or consortia of microorganisms, and ZRA and potential measurements made in conjunction with evaluations of microbiological activities. Current direction and magnitude can be indicative of the mechanisms and extents of MIC. The SC-ZRA setup mimics the heterogenous biofilm coverage of metal surfaces that leads to MIC, and we use it to establish the electrochemical signatures of MIC.

In this work, we use consortia and pure cultures of microorganisms isolated from samples taken in natural gas transmission and crude oil pipelines to inoculate SC-ZRA and incubate them under conditions representative of pipeline fluids. We specifically evaluated the impact of differential aeration, fluid flow, and type of microbial metabolism (fermentation, sulfate reduction, methanogenesis, and acetogenesis) on current magnitude, direction, and extent of corrosion. The identification of microbiological processes with electrochemical signatures can then be used to inform the implementation of ZRA measurements as a robust, sensitive and broadly applicable MIC monitoring approach.

611476 The Case of the Missing Asphaltenes from the Naphthenic Froth Treatment Process

Abstract:

Asphaltenes are a complex polydisperse system, and the molecular structure, kinetic and thermodynamic properties vary within a crude oil sample. The effects of this polydispersity can be manifested in various ways. For example, the sizes of asphaltene clusters follow a power-law distribution, where larger asphaltenes exist in decreasing likelihood[1] and the largest are nearest their phase transition[2]. Further, the double bonds equivalent of asphaltene molecules is inversely correlated to asphaltenes solubility[3], and surface-active asphaltenes can be separated by adsorption onto inorganic or water interfaces[4]. Therefore, the isolation and investigation of the most impactful asphaltenes are pivotal for understanding the influence of asphaltenes on different phenomena. In this presentation, we will present evidence that a fraction of asphaltenes is lost during the naphthenic froth treatment process used in the oil sands industry. We used toluene to dilute two bitumen samples, one extracted by toluene and the other one extracted by naphtha, then titrated heptane into diluted bitumen solution to destabilize asphaltenes. The naphtha extracted bitumen needs higher heptane concentration to destabilize asphaltenes, which implies the poorly soluble asphaltenes are lost in the naphthenic froth treatment process. These "missing asphaltenes" are hypothesized to migrate to water-in-oil and inorganic clay interfaces, and are thus removed from the bitumen product. The effect of these processes on stabilizing water-in-oil emulsions is of critical concern to improve product quality and reduce corrosion concerns. Evidence for this hypothesis comes from observations on asphaltenes and solids co-aggregation, where co-aggregation clusters were cut by Focused Ion Beam and imaged by SEM. Nano-scale (20~200 nm) clay particles were found to be distributed in the precipitated asphaltene particles. Based on these findings, our objective is to isolate and investigate the missing asphaltenes, which may be the "tail end" of the size distribution or the poorly soluble asphaltenes. Based on our preliminary results, approximately less than 1% of the asphaltenes (equivalent approx. 0.1% of bitumen mass) are lost in the naphthenic bitumen extraction. To extract and characterize this "missing" fraction, we assembled a nano-filtration system to isolate the most unstable asphaltenes and the nano-scale inorganic solids these asphaltenes may associate with. The filtration process was operated under process conditions with a bitumen (B) to diluent (D, naphtha) ratio of D:B = 0.7 at 80 °C.

Reference

[1] Fenistein, D., and L. Barre. Fuel 80.2 (2001): 283-287.

[2] Zhao, Bei, and John M. Shaw. Energy & Fuels 21.5 (2007): 2795-2804.

[3] Clingenpeel, Amy C., et al. Energy & Fuels 31.6 (2017): 5933-5939.

[4] Jarvis, Jacqueline M., et al. Energy & Fuels 29.11 (2015): 7058-7064.

607067 Advancements in Laboratory Testing for Effective Gas Hydrates Management in the Flow Assurance of Oil and Gas Production

Abstract:

Gas hydrates is a significant part in the portfolio of topics covered in the flow assurance of oil and gas production. Major advances have been made in understanding hydrates formation in multiphase system, which has enabled a gradual shift in practice to hydrate management from avoidance. In hydrate management, a solid understanding is required from the formation conditions (thermodynamics) to the formation process (kinetics) to the transportability (multiphase flow) in flowlines. As such, it is important for laboratory testing to provide the essential information needed to develop quantifiable metrics for the risks associated with hydrate management.

A number of testing equipment are typically used for testing of hydrate formation in the laboratory. The conventional testing tools include high pressure autoclave, high pressure rheometer, and rocking cells, in addition to fluid characterization such as wetting index, emulsion stability, and chemical composition. More recently, we have developed a number of different tools and techniques to broaden our insight and reliability of the knowledge to be translated from the laboratory to the field. The tools developed include a single methodology for accurate hydrate equilibrium boundary measurements and predictions in inhibited systems, an innovative setup for hydrate transportability in multiphase flow systems with a simple metric for flow risk, and a novel approach for modeling hydrates formation accounting for kinetics and agglomeration.

The tool for hydrate equilibrium for inhibited systems is based on the simple concept of depression temperature, which can be easily determined from water activity. We will demonstrate how this methodology works for simple and complex systems alike, eliminating the need for time-consuming and expensive measurements. The second tool is the rock-flow cell for hydrate formation under multiphase flow conditions, which is a setup capable of providing information to assess the agglomeration, deposition, and bedding of hydrates. Together, these results can be summarized and understood in terms of a hydrate flow risk index, which is a measure of the transportability of hydrate slurries. Finally, a new concept for hydrate formation based on dry vs wet is introduced to explain the actual mechanism for hydrate slurry transportability. Models for the hydrate kinetics and agglomeration using this concept are proposed with promising results.

The ideas and results presented will highlight both the simplicity and complexity of hydrates management in flow assurance. The goal of this presentation is to expose the community to recent developments in hydrate research that can significantly impact the understanding and implementation of management strategies that are technically effective and lower the risk in the decision making of flow assurance practices.

611242 Latest Advancements in Midstream Technologies

Abstract:

The advancements in horizontal drilling and hydraulic fracturing in the last couple of decades has shifted U.S. petroleum sector from imported to domestic production. Oil and gas (O&G) exports just from the state of Texas surged from $1.2 billion in 2006 to $13.4 billion in 2015. However, midstream infrastructure and the complete value chain have not been able to keep pace with these levels of growth in upstream capacity. In addition, ~ 40% of pipelines are 50 years old (or older), owing to increased risks of corrosion and failure. From natural gas flaring to produced water, critical environmental and economic considerations must be at the forefront of today's O&G industry for long term viability.

The Center for Midstream Management and Science at Lamar University has established a framework for advancing the midstream sector with 21st Century technologies. Current projects include increased resiliency of pipelines using drones for leak detection, distributed sensors for IIoT, and integration of the complete O&G supply chain. Additionally, innovations in produced water, both thermo-chemically and bio-chemically, decrease contaminants and increase waste-to-value near wellheads. An overview of these technologies will be presented as well as economic and infrastructure projections for the midstream sector.

597463 Quantifying Wetted Area during Multiphase Flow in an Oil Reservoir

Abstract:

Wettability of reservoir sediments is generally quantified using contact angle, which is importnt for reservoir characterization for enhanced oil recovery. Contact angle measurement, Amott Harvey index and USBM method are used to quantifying wettability. These methods does not account for heterogeneity of an oil reservoir and are performed at much smaller scale than an oil reservoir. In this work, we develop a method of quantifying wettability in a heterogeneous, large scale porous medium. We use dual tracer experiments to quantify the water and oil wetted surface area of a porous medium. We use two tracers for phase, one tracer being an ideal tracer and follows oil or water flow in the reservoir. Another tracer we use is an adsorbing tracer, which dynamically adsorbs and desorbs on the porous surface surface. We conduct the experiments at various saturations of the liquid and various wettability conditions. Using the contact area with the sediments vs. saturation curves we develop a method to quantify wettability in terms of the wetted area. The new quantification of wettability compares well with other quantification methods. This method can be done during the ongoing production from the reservoir and takes into account the heterogeneity of the porous surface.

605013 Impacts of Relaxation Time and Cohesive Energy Density on Syneresis Time of Hydrogels Under Harsh Reservoir Conditions

Abstract:

Hydrogels have been proven as promising treatment agent for water shutoff and conformance control applications. However, the biggest disadvantage of these materials is in their weak and brittle structures. Hence, improving the performance of hydrogels under harsh reservoir conditions is the objective of this study. Adopting the key thrust in this field, we attempt to optimize the energy dissipating mechanisms within the hydrogel network structure.

Our approach is to create hydrogels with a large number of dynamic network interactions through increasing the density of hydrogen bonds and, thus, improving the cohesive energy density (E_c). The oscillatory sweep and frequency sweep tests were carried out on a series of hydrogels of sulfonated polyacrylamide/chromium triacetate prepared using various initial concentrations of polymer and crosslinker, to derive a coherent hydrogel structure. The characteristic frequency and relaxation time of each hydrogel was determined, based on the modifications of the elastic, viscous and complex moduli. Finally, the performance of the optimal hydrogel was investigated by conducting a core-flooding test.

The cohesive energy density was calculated for all samples and compared with other parameters. A relationship between the cohesive energy density and the relaxation time was established. It was observed that as the cohesive energy density increases, the relaxation time of the network also increases. Our results suggest that the longer the relaxation time of a hydrogel, the longer its syneresis time will become. In some hydrogel networks, with cohesive energy densities in the range of 40´10^{7} to 3´10^{7} J/m^{3}, no volume reduction occurred over a period of three months. However, for another hydrogel sample with a cohesive energy density of 10^{4} J/m^{3} and the lowest relaxation time of all those samples examined in this study, syneresis was realized only after about 25 days. Based on the experimentally derived hydrogel structures, the optimal composition was identified, and its performance in reduction of water cut was evaluated by conducting a core-flooding test on a fractured core, in the presence of formation water with 200,000 ppm total salinity at 90°C. The optimal hydrogel sample retained its structural and viscoelastic behavior under a pressure differential of (at least) 100 bar while successfully reducing the produced water cut by 85%.

It is concluded that the successful performance of hydrogels under harsh reservoir conditions can be achieved by improving their viscoelastic structures. E_c is a valuable parameter to consider when investigating the structural strength of hydrogels. The performance and lifetime of hydrogels in reservoir environments can be significantly optimized by improving the rheological properties of hydrogels, E_c, and adjusting the relaxation time.

605342 Quantifying Uncertainty in Methane Hydrate Reservoir Simulations Using Monte Carlo Method

Abstract:

Methane gas hydrate deposits occurring in permafrost regions and the deep oceans are known to store a large volume of methane gas. The extraction of methane stored in these deposits can become an efficient replacement for the existing conventional energy resources. On the other hand, extraction of methane, which is a potent greenhouse gas, can reduce the chances of global climate change due to the spontaneous dissociation of gas hydrates. The short term production tests in the last decade clearly indicate the possibility of methane production from the gas hydrate deposits. However, the long term productions still have some geological and technical issues. Reservoir simulation is an extremely useful tool to forecast the production potential and to reduce the capital cost during field test, which require knowledge of geological and operational factors that control the methane production. However, these parameters, such as, porosity, saturation of hydrate, permeability, and initial temperature-pressure conditions are the major source of uncertainty in the reservoir model due to sparse knowledge about the reservoirs. This study aims to quantify uncertainty in simulation predictions for uncertainty in reservoir parameters using Monte Carlo method. We use class-2, confined, oceanic reservoirs.

An In-house multicomponent, multiphase, thermal, 3-D finite volume simulator is used that considers three components - water, methane and hydrate in four phases - gas, aqueous, hydrate and ice. The simulator has been validated against several other simulators in the code comparison study conducted by US DOE and history matched with hydrate formation and dissociation experiments. Energy and mass balance equations are solved in space and time domain to compute the production of gas in a reservoir. A confined aquifer layer below the hydrate zone is considered in this work where depressurization method provides a satisfactory results for gas production.

We consider a Gaussian distribution of porosity, initial hydrate saturation and initial temperature with 7.5% standard deviation from the mean. The results are obtained from 1000 deterministic simulations from each set of variable as a function of time. Total uncertainty in the gas production with $\pm 2\sigma$ bound calculated from Monte Carlo simulations. We see that gas production is most sensitive to porosity; however, initial temperature does not affect the production significantly. The mean production profile shows more dispersion and lower cumulative gas production than the deterministic simulations at mean input porosity and the initial hydrate saturation.

606173 An Improved Method for Predicting Autoignition Temperatures Based on First Principles

Abstract:

Methods for predicting autoignition temperatures (AIT) have been historically inaccurate and are rarely based on the underlying physical phenomena leading to observed AIT. Previous attempts at predicting AIT have been developed based on Quantitative Structure-Property Relationships (QSPR) and additive group contribution methods. This work presents an improved method for predicting AIT based on the method originally developed by the late Dr. William H. Seaton. The method of Seaton is described in detail including its use of first principles to predict AIT. Improvements on the method of Seaton and underlying principles are presented and discussed.

611821 Water Washing of Startup Crude Oil

Abstract:

During startup of cold sub-sea wells, methanol is injected at the wellhead to prevent the formation of hydrates in the pipeline/riser system. Methanol injection continues until the fluid arrival temperature exceeds hydrate-formation temperature at the operating pressure, and the typical duration is a few days. Methanol contamination is problematic, and hence the methanol specification of the crude oil is usually less than 20 ppmw. Water or brine wash is often used to reduce the methanol content of the startup crude oil. This paper studies the design of these systems using process simulation. The enabling need is for a reliable and accurate thermodynamic model. The model must describe the three-phase system containing the aqueous and oil liquid phases and the vapor phase, and the key needed feature is description of the distribution coefficient of methanol between the oil and aqueous phases. A model has been developed based upon the ElectrolyteNRTL model in Aspen Plus, and the presentation describes the key data sources and the parameter development. The model enables design of an optimum process flowsheet, and a two-stage countercurrent has been determined to be the optimum option.

607019 Universal Correlation for the Binary Interaction Parameter of Hydrogen with Petroleum Fractions in Refinery Hydroprocessing Applications

Abstract:

Hydrotreating and hydrocracking applications in oil refineries use large quantities of hydrogen. Hydrotreating removes hetero-atoms of sulfur, nitrogen and oxygen and can saturate aromatic ring structures, producing more desirable transportation fuels, such as ultra-low sulfur diesel. Hydrocracking is a more severe reaction, which breaks carbon-carbon bonds and improves the ratio of hydrogen to carbon in the fluid, again, making more suitable transportation fuels. Hydrogen is expensive and is either made on-site, perhaps through steam reforming of methane, or is bought from outside as a utility.

To design new hydroprocessing facilities and to improve the performance of existing plant, it is essential to be able to predict the solubility of hydrogen in typical hydrocarbon fluids. The hydrogen has to be separated from the product streams in high and low pressure separators and is then recycled to the reactor feed, where it is combined with hydrogen make-up.

Although Detailed Hydrocarbon Analysis (DHA) enables us to define the composition of reactor feedstocks in terms of a larger slate of defined chemical components, it is still common practice to represent the heavier portion of the feed in terms of undefined components, or petro fractions, characterized by narrow cut boiling points and densities from a petroleum assay.

Cubic equations of state (EoS), such as Soave-Redlich-Kwong and Peng-Robinson are the workhorse thermodynamic methods for calculating vapor liquid equilibrium (VLE) in refinery processes. From the study of the VLE of binary systems of hydrogen as a solute in a defined hydrocarbon solvent, it is evident that a significant correction to the basic Van der Waals mixing rule in the EoS is required if we are to predict accurately the composition of hydrogen in the liquid phase. Moreover, we can show that this correction, through a binary interaction parameter (k_{ij}) needs to be temperature-dependent. While we can correlate k_{ij} for hydrogen binaries with defined components through regression of experimentally measured VLE data, the problem is how we can predict the interaction between hydrogen and any petro fraction? We observe that increasingly large values of k_{ij} are required as the molecular weight of the hydrocarbon increases and any petro fractions are likely to be in the higher molecular weight range. Ignoring the interaction parameter between hydrogen and a petro fraction will result in significant inaccuracy in any prediction of solubility.

In this study, we will show how the wealth of binary VLE data for hydrogen with defined hydrocarbon solvents can be reduced through regression and conversion to Henry's Law constants that correlate with the reduced temperature and critical pressure of the solvent [1,2]. We propose a new universal correlation for these Henry's Law constants that will allow the subsequent prediction for k_{ij} between hydrogen and a petro fraction, based only on knowledge of the critical temperature, critical pressure and acentric factor of the petro fraction that can be calculated from well-established estimation methods. The validity of this approach will be shown by comparing the predicted solubility of hydrogen in a well-analyzed diesel stream with a base case using methods currently described in the Technical Data Book of the American Petroleum Institute [3].

[1] Trinh TKH, de Hemptinne JC, Lugo R, Ferrando N, Passarello JP, "Hydrogen Solubility in Hydrocarbon and Oxygenated Organic Compounds." J. Chem. Eng. Data 2016,61,119-34

[2] Fernández-Prini R, Alvarez JL, Harvey AH, "Henry's Constants and Vapor–Liquid Distribution Constants for Gaseous Solutes in and at High Temperatures "J. Phys. Chem. Ref. Data, 3012, 32, 903-912

[3] API Tech Data Book 10, 2019

600821 Distillation Curve and Vapor Pressure Predictions of Alcohol-Gasoline Mixtures Using an Advanced Equation of State

Abstract:

Due to federal environmental regulations, such as the Energy Policy Act of 2005 in the United States, ethanol is frequently blended into gasoline in an effort to reduce tailpipe emissions. Moreover, there is a shift towards increasing the use of more sustainable energy resources such as biofuels, which includes ethanol, as transportation fuels. A fundamental understanding of how blending alcohol into a gasoline stream affects the properties of said stream is of critical importance. Specifically, the change in the distillation curve and vapor pressure is essential from a manufacturing and engine performance perspective. In the plant setting, there exists empirical correlations to predict these changes. However, these are typically very limited in scope as a correlation is only applicable to a specific plant gasoline mixed with a specific alcohol in a certain blend ratio.

Process simulators of petroleum process rely on equation of state (EoS) predictions of the phase behavior of petroleum pseudo-components defined by normal boiling point T_b, specific gravity *SG*, and *MW*. For gasoline blending these pseudo-components are typically generated from D86 curves. Hence, any method to be widely used in the prediction of alcohol / gasoline blending in refining processes should be accurate for petroleum pseudo-components.

In this work, we present a fundamental methodology for alcohol / gasoline blend property prediction using the polar PC-SAFT EoS[1] combined with the recently developed petroleum parameterization methodology, TAPPS (formerly referred to as EMPETRO)[2-6]. TAPPS is a highly accurate methodology used to determine hydrocarbon PC-SAFT parameters for petroleum species defined solely by T_b, *SG*, and *MW*. In addition to highly accurate predictions for pure component vapor pressures and densities, TAPPS accounts for polar features of petroleum such as the fact that aromatic species receive hydrogen bonds. This makes it well suited to describe alcohol / gasoline blending.

The approach presented here employs the base gasoline D86, average gasoline SG, and dry vapor pressure equivalent (DVPE) to generate gasoline pseudo-components. Using these pseudo-components, it is demonstrated the PC-SAFT / TAPPS approach allows for the quantitative prediction of the D86 and DVPE of alcohol / gasoline mixtures. We validate the approach by reproducing both literature[7,8] and ExxonMobil refinery data[9] for various blends. This work demonstrates the strength of PC-SAFT / TAPPS in predicting the physical properties of non-traditional blends of petroleum streams and lays the foundation for the development of a predictive blending tool that will be valuable when considering emerging biofuel markets.

References

[1] J. Gross; G. Sadowski, *Ind. Eng. Chem. Res.*, **40** (2001) *1244.*

[2] B. D. Marshall; C. P. Bokis, *Fluid Ph. Equilibria*, **489** (2019) *83.*

[3] B. D. Marshall; C. P. Bokis, *Fluid Ph. Equilibria*, **478** (2018) *34.*

[4] B. D. Marshall, *Fluid Ph. Equilibria*, **493** (2019) *153.*

[5] B. D. Marshall, *Fluid Ph. Equilibria*, **497** (2019) *79.*

[6] B. D. Marshall, *Fluid Ph. Equilibria*, **507** (2020) *112420.*

[7] V. F. Andersen; J. E. Andersen; T. J. Wallington; S. A. Mueller; O. J. Nielsen, *Energy Fuels,* **24** (2010) *2683.*

[8] V. F. Andersen; J. E. Andersen; T. J. Wallington; S. A. Mueller; O. J. Nielsen, *Energy Fuels,* **24** (2010) *3647.*

[9] J. R. Vella; B. D. Marshall, *Ind. Eng. Chem. Res.* (2020) *accepted.*

606540 Development of Mean Ionic Activity Coefficients for Modification of the Krichevsky – Ilinskeya Equation for Methylimidazolim – Type Ionic Liquids

Abstract:

Removal of H_2S from oil and gas streams has been a priority, oftentimes a necessity, for upstream crude oil/gas facilities for decades. However, the advances in production, namely horizontal drilling and hydraulic fracking, have produced oils and natural gases that contain higher concentrations of H_2S than previously seen. Crude oils containing > 0.5 % sulfur are considered sour and must undergo separation and/or conversion processes to lower the H_2S content prior to entering the midstream pipelines and railcars. H_2S scavengers can be categorized into two types: non-regenerative and regenerative. Non-regenerative scavengers, such as triazines, use an irreversible reaction to convert H_2S forming sulfur-containing hydrocarbons. With regenerative scavengers, such as alkanolamines, absorb H_2S using reversible absorption with stripping recovery processes.

Ionic Liquids (ILs) are a new family of absorbent materials that are being sought to replace amine-type reversible absorbents due to lower production of volatile organic carbons (VOCs) and reduced costs of the regenerative stripping phase. The ILs that have been chosen for this work were 1-butyl-3-methylimidazolium hexafluorophosphate, 1-butyl-3-methylimidazolium chloride, and 1-butyl-3-methylimida-zolium tetrafluoroborate. In this work, the mean ionic activity coefficient and the adjustable parameters for the Pitzer Ion Interaction Model have been determined. An infinite dilution procedure has been employed where a concentrated mixture of H_2S in IL was stripped using helium gas. The decrease in H_2S concentration (measured through gas chromatography) over time was used to determine the pertinent thermodynamic parameters.

In addition, the Henry's constant and partial molar volume of H_2S in IL were determined. The Krichevsky – Kasarnovsky equation has been shown to be applicable over a wide range of temperatures, pressures, and solute/solvent systems. However, in some cases (i.e. where T is near T_c) the molar volume ( ) is not constant, and there are strong deviations between the measured solubility and the calculated Henry's constants. From preliminary estimates, the Krichevsky – Ilinskeya equation (a modification of the Krichevsky – Kasarnovsky equation) does not model the phenomenon well. Ultimately, this work seeks to establish a new thermodynamic model that replaces the 2-suffix Margules component of the Krichevsky – Ilinskeya equation with the mean ionic activity coefficient found in Pitzer Ion Interaction model. The experimental results and model development will be presented.

609189 3D Vof Simulation of Intermittent Flows in Inclined Pipes

Abstract:

Slug flow is one of the most predominant and unwanted flow pattern in multiphase oil and gas pipelines, severe slugging knocks off the separation systems. Accurate prediction of slug formation and slugging frequency leads to a better design and operation of oil and gas separation units. The multi-dimensional nature of fluid dynamic process that characterises slug flow as well as its transient nature, has made it more difficult to fully understand. However, recent advances in computational power has made Computational Fluid Dynamics (CFD) one of the key tools for investigating intricate flow behaviour. In this research, The Volume of Fluid (VOF) model of ANSYS Fluent was used to carry out a set of simulation runs with different liquids and gases in three different pipes with internal diameters of 20, 40, and 60 mm, across a superficial velocity ranges of 0 to 2 m/s for the liquid phase and also 0 to 20 m/s for the gas phase. To study the combined effects of viscosity, velocity and inclination angle, five different pipe orientations was investigated; 30 degrees upward inclined, 15 degrees upward inclined, horizontal pipe, 15 degrees downward inclined, and 30 degrees downward inclined. Some of the simulation runs are verified experimentally, and there is a good agreement between the simulation output and experimental results.

610054 Effects of the Subphase pH on the Isothermal Compressibility of Interfacial Molecular Films Containing Crude Oil Amphiphilics

Abstract:

Petroleum amphiphilics present in the SARA fractions have the ability to aggregate in solution, adsorb on interfaces, and reduce the solvent interfacial tension. The aggregation phenomena have also been shown to be responsible for problems caused by phase separation and solid deposition of oil components, causing damage to the oil pipeline and the formation-rock. The adsorption of crude oil amphiphilics on oil-water and oil-air interfaces is responsible for the many oilfied problems. The asphaltenes and resins are able to adsorb onto interfaces, constructing resistant films with viscoelastic characteristics that contribute to the stability of crude oil emulsions. The properties of the molecular film adsorbed on the interface are crucial to define the main aspects of oil dispersion and wetting phenomena. The description of rheological and interfacial properties of interfacial films constituted of oil amphiphilics is of great relevance for oil recovery and production activities.

Interfacial phenomena are present in many operations of the petroleum production and transportation processes. Petroleum components have been extensively investigated with regard to their colloidal characteristics because of the effects on the emulsion and foam stabilization, wettability inversion, solid deposition, and adhesion on solid surfaces. The present work presents a study on the properties of amphiphilic components extracted from petroleum at water-air interfaces. The investigation describes the phase behavior of monolayers containing SARA fractions. The results obtained from Langmuir trough experiments identified the phase transition in saturates, aromatics, resins, and asphaltene films. The description of the phase behavior of asphaltenes films should contribute to the assessment of the rheological behavior and the stability of petroleum emulsions. Asphaltene films displayed the highest surface pressure, reaching 70 $mN.m^{-1}$ at the minima area, according to previous values reported[1]. Asphaltene and resin films show a continuous increase in surface pressure as a function of pH from the first phase transition. On the other hand, a minimum point in surface pressure is found in aromatic and saturated films at pH 6. The reduction of the modulus of elasticity with the increase in pressure denotes the substantial intermolecular interaction of molecules present in the fractions. Factors such as ionization and temperature play an important role in determining the nature of the film. The pH effect on the isothermal compressibility is related to the electric double-layer forces acting in the air-water interface. Isothermal compressibility of asphaltene films was found from 0.012 (LC) to 0.158 (LE) under the full pH range. The highest isothermal compressibility was found to saturate films (3.47) at pH 12. The ionization, dissolution and/or ion adsorption phenomena have a great influence on the distribution of ions in the polar environment. The occurrence of ionizable groups on the amphiphilic molecular structure is the major responsible by the changes of the film elasticity with the pH changes.

610112 Synergy of Crude Oil Amphiphiles on the Interfacial Film Elasticity

Abstract:

Asphaltenes and resins comprise the most polar components and a portion of the most interfacially active compounds in crude oil. Crude oil asphaltenes represent a solubility class formed by a series of different amphiphilic species. Asphaltene constitutes the fraction of the crude oil composed of brown and black composites, which are insoluble in low boiling n-alkanes and quite soluble in toluene, benzene, carbon disulfide, chloroform and other chlorinated hydrocarbon solvents.

In solution, asphaltenes reduce the interface tension of organic solvents, promoting foam and emulsion stabilization. In addition, asphaltenes display a self-association behavior and adsorb at interfaces similar to conventional surfactants. Asphaltene deposition is a major problem in oil production. The deposition can result in the reduction of reservoir-rock permeability and a decrease in the pipe effective diameter. Despite the variety of techniques and methodologies available to evaluate asphaltene deposition inhibition, the addition of dispersants and inhibitors has been the preferred method. Asphaltenes adsorb at the air-water and oil-water interfaces, forming monomolecular layers as a Langmuir film. The properties of the interfacial film enclosing dispersed droplets in petroleum emulsions are mainly dependent on the concentration of polar molecules in the oil, the composition and pH of the aqueous phase, the temperature and film compressibility. Strong films, which are responsible for stabilizing oil emulsions against coalescence, consist essentially of asphaltenes. Films containing primarily resins are weak and tend to form unstable emulsions. Furthermore, the characteristics of the asphaltene fraction affect the interfacial film properties. Asphaltene fractions that have low aromaticity and a high content of heavy metals (mainly Ni and V) have been noted as the major component responsible for constructing films with high elasticity and high yield stress, whereas more aromatic asphaltenes possessing low heavy metal content produce weak films.

In this work, the effect of asphaltenes on the interfacial oil properties is investigated regarding the elasticity of interfacial films. N-pentane insoluble asphaltenes (C5I) were extracted from a heavy crude oil by means of the IP143 method. Compression isotherms obtained for the oil and its asphaltene fraction were compared. The samples were deposited at the air-water interface from a dichloromethane spreading solution to form two-dimensional monolayers. The film properties were evaluated using a Langmuir trough under compression at a constant rate of 10 $mm.min^{-1}$. Surface pressure-area isotherms were interpreted from a mathematic model of the surface pressure, assuming constant compressibility for each phase, such as previous work[1]. The C5I monolayers displayed an extensive region containing liquid-expanded (LE) and liquid-condensed (LC) phases and a well-identified transition region in between these phases. Tests performed for oil shows a long gas region undergoing a phase transition at 0.20 $m^2.mg^{-1}$ and 1.83 $mN.m^{-1}$. The compression of the film produced a liquid expanded region to about 0.067 $m^2.mg^{-1}$ and 14.18 $mN.m^{-1}$. From that point on, a second phase transition occurs at 0.038 $m^2.mg^{-1}$ and 36.36 $mN.m^{-1}$ which corresponds to the maximum surface pressure. The interfacial behavior of the film containing C5I asphaltenes shows a phase transition to expanded liquid with a low surface pressure of 9.14 $mN.m^{-1}$, following a rapid phase transition to condensed liquid at 0.178 $m^2.mg^{-1}$ and 50.56 $mN.m^{-1}$ which is maintained up to the maximum pressure of 72.23 $mN.m^{-1}$, where the area corresponds to the minimum area of 0.039 $m^2.mg^{-1}$. Asphaltene films display a very large hysteresis area, which is related to irreversible strong interactions between the molecules adsorbed on the air-liquid interface.

CHARACTERIZATION OF FACTORS AFFECTING THE GAS WELL DELIVERABILITY OF THE JIAOSHIBA SHALE GAS FIELD BY MULTIVARIATE STATISTICAL ANALYSIS

Yizhong Zhang[1,2], Maolin Zhang[1], Bin Ju[3], Jincai Shen[4], Lan Ge[4], Fanhua Zeng[2]*

**Corresponding author. Email address: Fanhua.Zeng@uregina.ca.*

[1] Hubei Cooperative Innovation Center of Unconventional Oil and Gas, Yangtze University, Hubei, Wuhan, China, 430100

[2] University of Regina, Regina, SK, Canada, S4S 0A2

[3] School of Oil and Natural Gas Engineering, Southwest Petroleum University, Chengdu, Sichuan, China, 610500

[4] SINOPEC Chongqing Fuling Shale Gas Exploration and Development Co., Ltd., Chongqing, China, 408014

Abstract

The Longmaxi shale gas in the Jiaoshiba Shale Gas Field is the first and the largest successfully developed shale gas in China. The analysis of key factors of deliverability of gas wells plays a significant role in efficiently developing shale gas reservoirs. However, there are few methods analyzing the key factors of deliverability of shale gas wells by considering both geological characters and engineering parameters. In this study, the reservoir characteristics, drilling and completion data, fracturing data (including the well logging interpretation of horizontal section, the track and layer section that the well bore passed through, and fracturing parameters) and production data of 86 gas wells in the Wufeng-Longmaxi shale, Jiaoshiba Shale Gas Field in Eastern Sichuan Basin, China are statistically analyzed and ordered by multivariate statistical analysis. The gas well deliverability is described by absolute open flow potential and 9 factors (physical parameters of reservoir, including density of rock, porosity, total organic carbon, brittle mineral content of rock and drilling and fracturing engineering quality, including horizontal length of gas well, horizontal length ratio through different stages in the formation, number of fracturing stages, liquid volume, and sand amount) that may affect the deliverability of shale gas wells are considered. The results reveal that the absolute open flow potential has a close relationship with porosity, total organic carbon, brittle mineral content and total liquid volume. The multivariable nonlinear regression models of absolute open flow potential with different key factors considered are built, and the prediction results are reliable by goodness-of-fit test acceptable with real gas well deliverability testing data. The research methods and results in this study can provide a reference basis for reasonable production rate allocation, design, and adjustment of development plan for the shale gas reservoirs.

Keywords: Gas well deliverability, Longmaxi shale gas, Jiaoshiba Shale Gas Field, Multivariate statistical analysis, Multivariable nonlinear regression

1. Introduction

Following the increased demand of petroleum energy and the developed science and technology, the industry nowadays has focused on the unconventional resources. Shale gas reservoirs are one of the most popular kind of unconventional resources with extremely low permeability, tight porous structure, large specific area and high heterogeneity, which have attracted attentions of researchers worldwide [1-3]. Shale gas has self-generated, self-contained, organic-rich, nanoscale pore structure, and other unusual characteristics and it could be considered as a source as well as a reservoir [4-5]. In addition to free gas and small amount of dissolved gas, adsorbed gas is largely existed in the tight formation, which makes it difficult to produce owing to the complexity of the reservoir [6-8]. The Sichuan-Chongqing region in southwest China is one of the fastest shale gas production growing regions in the world. The annual shale gas deliverability of China has reached 2×10^{10} m^3 with the annual shale gas production of 1.5×10^{10} m^3 [9]. As one of the main contributors of natural gas production growth in China, the gas production of Jiaoshiba shale gas reservoir has exceeded the daily production of 1.67×10^7 m^3/d since 2013 [10]. The large production of commercial shale gas flow of the Longmaxi Formation (e.g. more than 2×10^5 m^3/d of JY1 well) has made the Jiaoshiba area one of the most important shale gas field in China.

Compared with typical shale gas reservoirs in North America, the shale gas in Sichuan Basin has lower TOC value, lower porosity, lower gas saturation, smaller thickness and higher buried depth, which makes its exploration more complicated [9]. In both the early test stage and current production stage, there are large differences among deliverability of the shale gas wells in the Jiaoshiba area. Therefore, the deliverability evaluation is the basis of reasonable shale gas production and prediction, and the research on deliverability evaluation and its related reservoir and engineering parameters of shale gas wells is particularly important. Previous studies have focused on the impacts of reservoir quality including sedimentary environment [11], structure [12-13], mineralogy and lithology [14-17] and engineering quality including drilling and completion techniques [12, 18] and multi-stage hydraulic fracturing performance [19-20] on the shale gas production. Their results revealed that both the reservoir quality and engineering quality will affect the produced gas rate of the shale gas wells. Most of researches did analysis based on the relationship between the interested variable and its single impacting factor, comprehensive multivariate statistical analysis with consideration of correlation and collinearity among factors needs to be further studied. Moreover, instead of the produced gas rate or gas bearing content, the absolute open flow (AOF) potential may be a better parameter to describe the deliverability of gas well. The AOF potential is a hypothetical gas flow rate that may be produced if the flowing pressure at the bottom of the well (BHFP) were equivalent to the atmospheric pressure, i.e. the maximum flow rate that the gas well may produce [21]. In this study, the well deliverability of Jiaoshiba shale gas is described by the AOF potential based on BHFP and the impacts of reservoir and engineering characteristics on the AOF potential will be comprehensively analyzed. The data of over 86 wells of Fuling Jiaoshiba shale gas field is analyzed and the impacts of the reservoir quality (i.e. density of rock, porosity, total organic carbon, brittle mineral content of rock) and the engineering quality (i.e. horizontal length of gas well, horizontal length ratio through different formation stages, number of fracturing stages, liquid volume, and sand amount) on the AOF are quantified. The most important factors affecting deliverability of the gas wells in the Jiaoshiba area will be obtained by hierarchical clustering analysis method. The regression correlation between the well deliverability and the important parameters will be built. The objective of this study is to find a new method to evaluate the key factors that affect the well deliverability of shale gas reservoir by multivariate statistical analysis with consideration of correlation and collinearity among factors and build multivariable nonlinear regression models based on these key factors, so that the well deliverability of shale gas reservoir could be predicted.

2. Shale gas reservoir characters

The Longmaxi shale gas in the Jiaoshiba area located in the Eastern Sichuan Basin in the southwest region of China is the first and the largest successfully developed shale gas reservoir in China [22]. The shale gas bearing formation is the Upper Ordovician Wufeng Formation to Lower Silurian Longmaxi Formation with a thickness of about 200–300 m, area of 272.44 km^2 and buried depth of less than 3500m. Comprehensively considering the lithology, electrical and physical properties, geochemical and gas-bearing characteristics, the gas bearing shale section in the Upper Ordovician Wufeng Formation to Lower Silurian Longmaxi Formation could be vertically divided into three parts with stable horizontal stratigraphic distributions for each member (i.e. Long1, Long2 and Long3 from bottom to top). Long1 member is rich in fossils (e.g., radiolarian, calthrop, and graptolite) and is composed of black mudstone and argillaceous siltstone. The Long2 member consists of argillaceous siltstone and fine-grained sandstone. The Long3 member is composed of dark-gray mudstone and gray shale with fewer fossils [5, 23].

Previous studies suggested that Long1 member consisted of one 3-ord sequence, and could be divided into three system tracts (i.e. the transgressive system tract (TST), early highstand system tract (EHST), and late highstand system tract (LHST)) [10, 15, 16]. As shown in **Fig.1** [5], this study is targeted in the Long1 member, which is abundant of organic matter with pores and could be divided into 3 stages (i.e. stage1, stage2 and stage3 from bottom to top) based on physical properties with gas bearing increasing from top to bottom. The Long1 member is of good brittleness, and the brittle mineral content gradually increases from top to bottom. The natural gas in the layer belongs to over-mature dry gas, in which methane is the main component without containing hydrogen sulfide, carbon monoxide and hydrogen.

3. Samples

The AOF potential is a hypothetical gas flow rate that may be produced if the flowing pressure at the bottom of the well (BHFP) were equivalent to the atmospheric pressure [21]. The AOF potential may be a better parameter to describe the deliverability of the gas wells as it gives the maximum flow rate that the gas well can produce. As shown in Fig.2, the AOF potential data of and 9 factors that may affect the deliverability of gas wells (i.e. horizontal length, number of fracture stages, total liquid volume, total amount of sand, rock density, porosity, TOC, brittle mineral content of rocks, horizontal length ratio through stage 1 and stage 3 of Long1 member) are collected (data is offered by SINOPEC Chongqing Fuling Shale Gas Exploration and Development Co., Ltd., Chongqing, China). Though **Fig.2** shows certain relationship between AOF potential and single impacting factor, the correlation and collinearity among different factors are not considered. According to the data, the average reservoir pressure of the Long1 member is 37.69MPa with pressure coefficient 1.55. The reservoir temperature is 85.99°C. The pay zone thickness is between 20m-200m and the most favorable thickness is about 38m. The kerogen type of the organic matter is mostly type II_1 to II_2. The total organic carbon (TOC) contents of the shale range from 1%-8% with average value of 5% and it is increasing from top to bottom. The vitrinite reflectance (R_o %) of organic matter ranges from 2.5%-3.7% with average value of 2.9%. The porosity of the shale is between 1.88% to 7.06%, with average value of 4.15% and the permeability of the shale is less than 0.001mD. The horizontal wells in the sample have lateral length between 1500m to 1800m, well spacing between 500m to 600m, and the distance between adjacent clusters between 20m to 25m. All the horizontal wells are multi-stage hydraulic fractured with number of fracturing stages between 15 to 29. The fluid volume per stage is between 1500m^3 to 1800m^3, and proppant volume per stage is between 50m^3-65m^3. The fractured half-lengths range from 300m to 350m, and propped half-lengths range from 250m to 270m. The fracture heights range from 40m to 50m, and main fracture width range from 5mm to 10mm.

4. Methods

4.1. Normality test

When working with multivariable regression problems, some assumptions should be followed. The regression process of independent variables and the target variable are assumed to follow Gauss–Markov theorem (i.e. the ordinary least squares (OLS) estimator has the lowest sampling variance within the class of linear unbiased estimators, if the independent variables have no perfect collinearity, and the errors in the regression model are uncorrelated with equal variances and expectation value of zero) [24]. Thus, the first step of the multivariate statistical analysis is to test whether the sample variables follow normal distribution. The Q–Q (quantile-quantile) plot is applied, which is a probability plot that shows a graphical comparison between the samples and normal distribution by plotting their quantiles against each other [25]. If the points in the Q–Q plot approximately lie on the 45-degree line (i.e. $y=x$), the distribution of the sample being compared are similar enough to normal distribution. If normality test of a variable behaves not well, certain transformation (i.e. logistic transformation, logarithm transformation, index transformation, power transformation, etc.) of the variable and nonlinear regression model should be considered.

4.2. Standardization

Before the key factor analysis, the calculation of correlations between each parameter is essential. Since the units, scales and ranges of the factors are different, the first step is to standardize all the parameters based on the normality test results. The method applied in the study is the Fisher Z-transformation method. The standard score (Z score) for each data point is its difference from sample mean divided by the sample standard deviation.

$$Z = \frac{X - \overline{X}}{s} \tag{1}$$

where

X is the data point.
$\overline{X}$ is the sample mean.
s is the sample standard deviation.

4.3. Correlation analysis

After standardization, the correlations between each factor are calculated. This is to eliminate the repeated consideration of correlation and collinearity in the later multivariable regression. In other words, if two variables are strongly related, only the principle variable between them is considered in later analysis. The sample data has a number of n wells, and each well has a number of k variables including a dependent variable (X_k) which is Z scores for AOF potential and $k-1$ independent variables (X_1, X_2, ..., X_{k-1}) which are Z scores for the factors affecting AOF potential. The sample data could be shown in matrix form as follows

$$X = \begin{bmatrix} x_{1,1} & x_{1,2} & \cdots & x_{1,k} \\ x_{2,1} & x_{2,2} & \cdots & x_{2,k} \\ \cdots & \cdots & & \cdots \\ x_{n,1} & x_{n,2} & \cdots & x_{n,k} \end{bmatrix} \quad (2)$$

Then the correlation matrix for all variables would be

$$R = \begin{bmatrix} r_{11} & r_{12} & \cdots & r_{1k} \\ r_{21} & r_{22} & \cdots & r_{2k} \\ \cdots & \cdots & \cdots & \cdots \\ r_{k1} & r_{k2} & \cdots & r_{kk} \end{bmatrix} \quad (3)$$

$r_{ij}(i,j=1,2,...,k)$ is the correlation coefficient for $x_{ij}(i=1,2,...,k)$ and $x_{ij}(i=1,2,...,k)$, and the calculation equation is

$$r_{ij} = \frac{\sum_{m=1}^{k}(x_{im}-\overline{x_i})(x_{jm}-\overline{x_j})}{\sqrt{\sum_{m=1}^{k}(x_{im}-\overline{x_i})^2\sum_{l=1}^{k}(x_{jl}-\overline{x_j})^2}} \quad (4)$$

The value of correlation coefficient should be between -1 and 1. If the correlation coefficient is close to 1, the involving two variables have a strong positive relationship, and if it is close to -1, the involving two variables have a strong negative relationship. If the correlation coefficient is close to 0, this suggests that the involving two variables may be independent to each other.

4.4. Hierarchical clustering analysis

As long as correlations among factors are obtained, clustering method could be applied to find the key factors while eliminate the factors with small effect on AOF potential or strong correlation with other factors. Hierarchical clustering analysis (HCA) is a method to find the hierarchy relationship among variables, so that the variables could be divided into clusters [26]. HCA is performed on the normalized sample data set of the 86 well data including AOF potentials and 9 different factors, using centroid linkage clustering (Eq.5) as linkage criteria. The method will try to link the two most related factors in the data set to form a new cluster at each step (**Fig.3** agglomerative hierarchical clustering). The new merged cluster is then considered as a variable in the relation map and agglomerative approach is again used. Finally, a tree-hierarchy structure with linkage lines that join all the factors would be built.

$$D_{uv}^2 = (\bar{X}_u - \bar{X}_v)'(\bar{X}_u - \bar{X}_v) \quad (5)$$

, where $\bar{X}_u$ is the centroid for cluster C_u. $\bar{X}_v$ is the centroid for cluster C_v.

4.5. Multivariable nonlinear regression

Once the main factors that affect the AOF potential are found, other factors are either unessential to AOF potential or have strong correlation to the main factors, which could be eliminated. The next step is to linearize the main factors, including logistic transformation, logarithm transformation, index transformation, power transformation, etc. Then, a set of multivariable linear regressions is carried out to find the optimal regression model. If there are k variables, the linear regression for the variables is

$$y_i = \alpha + \beta_1 X_{i1} + \ldots + \beta_k X_{ik} + \varepsilon_i = \alpha + \sum_{j=1}^{k} \beta_j X_{ij} + \varepsilon_i, i = 1,2,\ldots,n \quad (6)$$

Least squares could be used to find the intercept a and coefficients β_j (j=1,2,...,k), which is to find the optimums for a and β_j (j=1,2,...,k) when the sum of squared residuals, *SSe*, is minimized. The squared residual r_i^2 is defined as the difference between real value and predicted value of the dependent variable.

$$r_i = y_i - \hat{y}_i = y_i - f(x_i, \alpha, \beta) \quad (7)$$

$$SSe = \sum_{i=1}^{n} r_i^2 \quad (8)$$

$\hat{y}_i$ is the estimation value of dependent variable y_i. $a, b_j (j = 1,2,\ldots,k)$ are the estimation values of intercept and coefficients $\alpha, \beta_j (j = 1,2,\ldots,k)$. Let the partial derivatives with respect to $a, b_j (j = 1,2,\ldots,k)$ equals 0.

$$\begin{cases} \dfrac{\partial SS_e}{\partial a} = 0 \\ \dfrac{\partial SS_e}{\partial b_j} = 0 \end{cases} \quad j = 1,2,\ldots,k \quad (9)$$

It could be also proved that $a, b_j (j = 1,2,\ldots,k)$ is the unbiased estimation of $\alpha, \beta_j (j = 1,2,\ldots,k)$.

5. Results and Discussions

5.1. Normality test

Fig.4 shows the Q–Q (quantile-quantile) plots for the 10 variables of the 86 wells. The results reveal that the points of most variables in the Q–Q plots approximately lie on the 45-degree line, which means they follow normal distribution well. The only variable may not behave with good normality is horizontal length ratio through stage 1 and stage 3 of Long1 member. This variable may need transformation (i.e. logistic transformation, log transformation, logistic transformation, etc.) in the later regression study. Thus, the multivariable nonlinear regression would be needed in this study.

5.2. Standardization and correlation of the variables

The matrix of correlation coefficients of all variables is shown in **Table 1**. **Table 1** suggests that there is no strong linear relation between the AOF potential and single one factor. Relatively, the AOF potential is positively related to the TOC, the brittle mineral content, and the total liquid volume. It is negatively related to rock density. Some factors show strong correlation with each other, and the number of fracture stages and total liquid volume have the highest correlation coefficient of 0.946. The drilling and fracturing engineering factors, including the number of fracturing stages, the length of horizontal well, the total liquid volume and the total sand amount, are highly linear related. Among the geological factors, the rock density is highly negative to TOC, brittle mineral content, and the TOC and the brittle mineral content of rock are highly linear related. The collinearity among some factors is very strong and may affect AOF potential modelling, which should be eliminated in the multivariate statistical analysis of AOF potential.

5.3. Hierarchical clustering analysis

According to **Table 1**, the number of fracturing stages and total liquid volume have the highest correlation coefficient, which is 0.946. These two factors could be firstly clustered together, and the central point of this cluster is calculated by centroid method. Then the correlation analysis is done again after applying the whole cluster as a variable. This progress is continued until all the variables converged to one cluster. The result of hierarchical clustering analysis with linkage lines is shown **Fig.5**, and **Table 2** gives the importance order of the effect of each factor on the AOF potential.

It is suggested in the hierarchical clustering analysis results that the order of the factors that affect AOF potential is: a. porosity, b. TOC, c. brittle mineral content of rock, d. horizontal length ratio through stage 1 and stage 3 of Long1 member, e. number of fracturing stages, f. total liquid volume, g. horizontal length, h. total sand amount and i. rock density. The AOF potential firstly has closest relationship to porosity. It then has closer relationship to the cluster of TOC, brittle mineral content of rock, horizontal length ratio through stage 1 and stage 3 of Long1 member. These four factors belong to the first cluster. The other four factors including number of fracturing stages, total liquid volume, horizontal length and total sand amount belong to the second cluster. Rock density belongs to the third cluster. Thus, the main factors that affect AOF potential are porosity, TOC, brittle mineral content of rock and number of fracturing stages.

5.4. Multivariable nonlinear regression

After the main clusters in the hierarchical clustering analysis are obtained, the key factors that affect the AOF potential are found. The independent variables in the multivariable nonlinear regression models could be chosen according to the cluster results. With consideration of the correlation and collinearity among factors, one factor in each cluster with smallest linkage distance is chosen in the regression model, and four models after testing (**Table 3**) are regressed in this study.

With trials of nonlinear transformation of the involving variables, the linear regression of the transformation is ideal, and all the coefficients have passed the student t test for significance. The four models are shown as follows.

Model 1:

$$Y' = \ln Y, X_1' = \ln X_1, X_2' = (\ln X_2)^8, X_3' = X_3 \tag{10}$$

$$Y' = -13.067 + 1.509X_1' + 0.012X_2' + 0.867X_3' \tag{11}$$

Model 2:

$$Y' = \ln Y, X_4' = \ln X_4, X_2' = (\ln X_2)^8, X_3' = \ln X_3 \tag{12}$$

$$Y' = -1.756 + 1.465X_4' + 0.011X_2' + 1.043X_2' \tag{13}$$

Model 3:

$$Y' = \ln Y, X_4' = \ln X_4, X_5' = X_5^{\frac{1}{100}}, X_2' = (\ln X_2)^8, X_3' = \ln X_3 \tag{14}$$

$$Y' = -1.603 + 1.434X_4' + 0.226X_5' + 0.011X_2' + 0.848X_3' \tag{15}$$

Model 4:

$$Y' = \ln Y, X_1' = \ln X_1^{\frac{1}{10}}, X_3' = X_3^{\frac{1}{100}}, X_2' = (\ln X_2)^8, X_6' = X_6 \tag{16}$$

$$Y' = -12.556 + 5.211X_1' + 0.149X_5' + 0.012X_2' + 0.035X_6 \tag{17}$$

It is shown from **Fig.6** that the total AOF potential of 86 wells is $5420.48\times10^4 m^3/d$ and the total AOF potential of 86 wells predicted by the 4 models range from $5110.24\times10^4 m^3/d$ to $5607.74\times10^4 m^3/d$. The average AOF potential of single well is $63.03\times10^4 m^3/d$ and the average AOF potential of single well predicted by the 4 models range from $59.42\times10^4 m^3/d$ to $65.21\times10^4 m^3/d$. The coefficient of determination (R^2) in the figure is the proportion of the variance in the dependent variable Y that is predictable from the independent variables $X_j (j=1,2,...,k)$. The R^2 values show that all the 4 models could be relatively preferable models to predict the AOF potentials based on different factors. In total, the prediction value of model 1 and model 4 are closer to the real AOF potential value with higher coefficients of determination (R^2) and less prediction errors.

6. Conclusions

In this study, instead of the produced gas rate, the absolute open flow (AOF) potential is analyzed as a better parameter to describe the deliverability of gas well. The reservoir parameters, drilling and completion data, fracturing data (including the well logging interpretation of horizontal section, the track and layer section that the well bore passed through, fracturing scale and monitoring parameters) and sample production data of 86 gas wells in the Wufeng-Longmaxi shale, Jiaoshiba shale gas field in Eastern Sichuan Basin, China are considered in the multivariate statistical analysis. The hierarchical clustering analysis could give the orders of the key factors that affect gas well AOF potential as well as eliminate the collinearity among the factors, which gives the basis of multivariable nonlinear regression. The results suggest that the order of the factors affects AOF potential is: a. porosity, b. TOC, c. brittle mineral content of rock, d. horizontal length ratio through stage 1 and stage 3 of Long1 member, e. number of fracturing stages, f. total liquid volume, g. horizontal length, h. total sand amount and i. rock density. The estimation results of the AOF potential by the multivariable nonlinear regression models based on the key factors from hierarchical clustering analysis have small differences to that of the actual values of the sample with goodness-of-fit test acceptable. The predictions of AOF potential of the models are reliable, and the feasibility of this multivariate statistical analysis is verified in the study of shale gas well deliverability. The research results can provide a reference basis for design of drilling and completion, the reasonable production rate allocation, design and adjustment of development plan of the shale gas reservoirs.

Acknowledgement

The authors would like to acknowledge the support from National Natural Science Foundation of China (NSFC Grant Number: 520004032) and the Thirteen-Five National Science and Technology Major Projects of China: 1. Study on Shale Gas Fracturing and Gas Production Regularity in Wufeng Formation and Longmaxi Formation in Fuling Area (Sub-Issue Number: 2016ZX05060-026) and 2. Research on the efficient development technology and policy of inter-salt shale oil reservoir (Sub-Issue Number: 2016ZX05049005-010).

References

1. Zolfaghari, A., Dehghanpour, H., & Xu, M. (2017b). Water sorption behaviour of gas shales: II. Pore size distribution. *International Journal of Coal Geology, 179*, 187-195.
2. Zhang, Y., Yao, S., Zhang, M., Zhou, X., Mei, H., & Zeng, F. (2018). Prediction of adsorption isotherms of multicomponent gas mixtures in tight porous media by the oil–gas-adsorption three-phase vacancy solution model. Energy & Fuels, 32(12), 12166-12173.
3. Zhang, Y., Zhang, M., Mei, H., & Zeng, F. (2019). Study on salt precipitation induced by formation brine flow and its effect on a high-salinity tight gas reservoir. *Journal of Petroleum Science and Engineering, 183*, 106384.
4. Dong, T., Harris, N. B., Ayranci, K., Twemlow, C. E., & Nassichuk, B. R. (2017). The impact of composition on pore throat size and permeability in high maturity shales: Middle and Upper Devonian Horn River Group, northeastern British Columbia, Canada. *Marine and Petroleum Geology*, 81, 220-236.
5. Wang, C., Zhang, B., Hu, Q., Shu, Z., Sun, M., & Bao, H. (2019). Laminae characteristics and influence on shale gas reservoir quality of lower Silurian Longmaxi Formation in the Jiaoshiba area of the Sichuan Basin, China. *Marine and Petroleum Geology*, *109*, 839-851.
6. Cao, T. T., Song, Z. G., Wang, S. B., & Jia, X. (2015). A comparative study of the specific surface area and pore structure of different shales and their kerogens. *Science China Earth Sciences*, *58*(4), 510-522.
7. Chalmers, G. R. L., & Bustin, R. M. (2007). The organic matter distribution and methane capacity of the lower cretaceous strata of northeastern British Columbia, Canada. *International Journal of Coal Geology, 70*(1-3), 223-239.
8. Chalmers, G. R. L., & Bustin, R. M. (2008). Lower cretaceous gas shales in northeastern British Columbia, Part II: Evaluation of regional potential gas resources. *Bulletin of*

Canadian Petroleum Geology, 56(1), 22-61.

9. Ma, X., Li, X., Liang, F. , Wan, Y. , & Guo, W. . (2020). Dominating factors on well productivity and development strategies optimization in Weiyuan shale gas play, Sichuan Basin, SW China. *Petroleum Exploration and Development, 47*(3), 594-602.
10. Shu, Y., Lu, Y., Chen, L., Wang, C., & Zhang, B. (2020). Factors influencing shale gas accumulation in the lower Silurian Longmaxi formation between the north and South Jiaoshiba area, Southeast Sichuan Basin, China. *Marine and Petroleum Geology, 111*, 905-917.
11. Wang, Y. , Dong, D. , Li, X. , Huang, J. , Wang, S. , & Wu, W. . (2015). Stratigraphic sequence and sedimentary characteristics of Lower Silurian Longmaxi Formation in the Sichuan Basin and its peripheral areas. Natural Gas Industry B, 2(2–3), 222-232.
12. Hu, M., Huang, W., & Li, J. (2018). Effects of structural characteristics on the productivity of shale gas wells: A case study on the Jiaoshiba Block in the Fuling shale gasfield, Sichuan Basin. *Natural Gas Industry B, 5*(2), 139-147.
13. Zhang, H., Wang, J., & Zhang, H. (2016). Investigation of the main factors during shale-gas production using grey relational analysis. *The Open Petroleum Engineering Journal, 9*(1).
14. Wei, C., Wang, H., Sun, S., Xiao, Y., Zhu, Y., & Qin, G. (2012, January). Potential investigation of shale gas reservoirs, Southern China. In *SPE Canadian Unconventional Resources Conference*. Society of Petroleum Engineers.
15. Wang, C., Zhang, B., Lu, Y., Shu, Z., Lu, Y., Bao, H., ... & Chen, L. (2018). Lithofacies distribution characteristics and its controlling factors of shale in Wufeng Formation-Member 1 of Longmaxi Formation in the Jiaoshiba area. *Petroleum Research, 3*(4), 306-319.
16. Wang, C., Zhang, B. Q., Lu, Y. C., Shu, Z. G., Lu, Y. Q., Bao, H. Y., ... & Chen, L. (2018). Lithofacies distribution characteristics and main development controlling factors of shale in Wufeng Formation-Member 1 of Longmaxi Formation in Jiaoshiba area. *Acta Pet. Sin., 39*(6), 631-644.
17. Kim, G., Lee, H., Chen, Z., Athichanagorn, S., & Shin, H. (2019). Effect of reservoir characteristics on the productivity and production forecasting of the Montney shale gas in Canada. *Journal of Petroleum Science and Engineering, 182*, 106276.
18. Curtis, M. E., Cardott, B. J., Sondergeld, C. H., & Rai, C. S. (2012). Development of organic porosity in the Woodford Shale with increasing thermal maturity. *International Journal of Coal Geology, 103*, 26-31.
19. Cipolla CL, Lolon EP, Erdle JC, Rubin B. Reservoir modeling in shale-gas reservoirs. *SPE reservoir evaluation & engineering*. 2010 Aug 1;13(04):638-53.
20. Zhu W, Ma D, Qi Q, Song Z, Yue M. Multi-zone coupling productivity of horizontal well fracturing with complex fracture networks in shale gas reservoirs. Natural Gas Industry B. 2018 Feb 1;5(1):80-9.
21. Mohammed, K. R., Stigliano, H., Vicente, M. G., & Angulo, R. J. (2010, January). Comparison of Different Deliverability Testing Techniques in Low Permeability Gas Reservoir and Its Impact on the Absolute Open Flow (AOF) Estimation. In *SPE EUROPEC/EAGE Annual Conference and Exhibition*. Society of Petroleum Engineers.
22. Guo, X., Qin, Z., Yang, R., Dong, T., He, S., Hao, F., ... & Liu, K. (2019). Comparison of

pore systems of clay-rich and silica-rich gas shales in the lower Silurian Longmaxi formation from the Jiaoshiba area in the eastern Sichuan Basin, China. *Marine and Petroleum Geology, 101*, 265-280.

23. Yang, R., He, S., Hu, Q., Hu, D., Zhang, S., & Yi, J. (2016). Pore characterization and methane sorption capacity of over-mature organic-rich Wufeng and Longmaxi shales in the southeast Sichuan Basin, China. Marine and Petroleum Geology, 77, 247-261.
24. Henderson, C. R. . (1975). Best linear unbiased estimation and prediction under a selection model. *Biometrics, 31*(2), 423-447.
25. Gnanadesikan, M. B. W. . (1968). Probability plotting methods for the analysis for the analysis of data. *Biometrika, 55*(1), 1-17.
26. Rokach, Lior, and Oded Maimon. (2005). "Clustering methods." Data mining and knowledge discovery handbook. Springer US, 321-352.

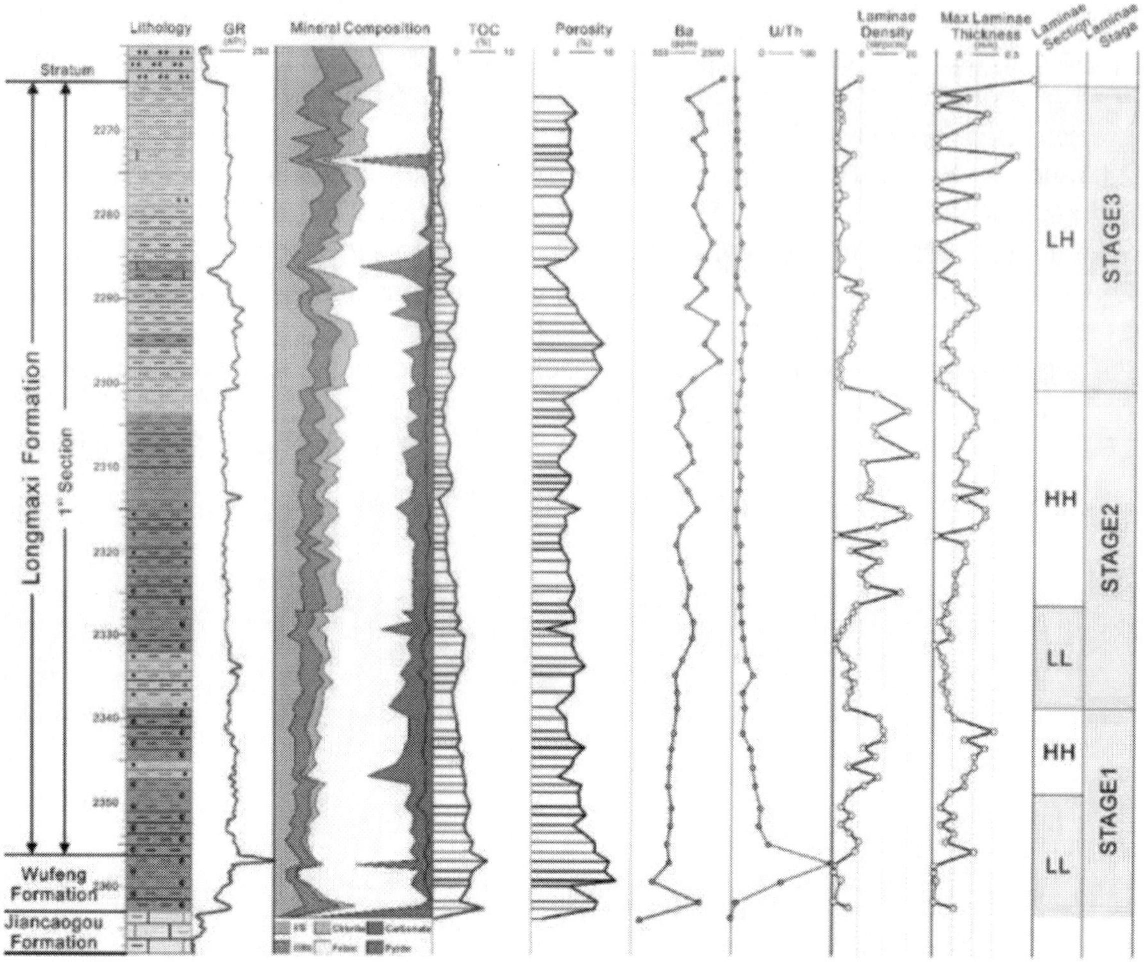

Fig.1 Lithofacies, gamma ray log, mineral content, TOC content, porosity, geochemical proxies, laminae density, and laminae stages of the Long1 member in well A (Wang et al., 2019).

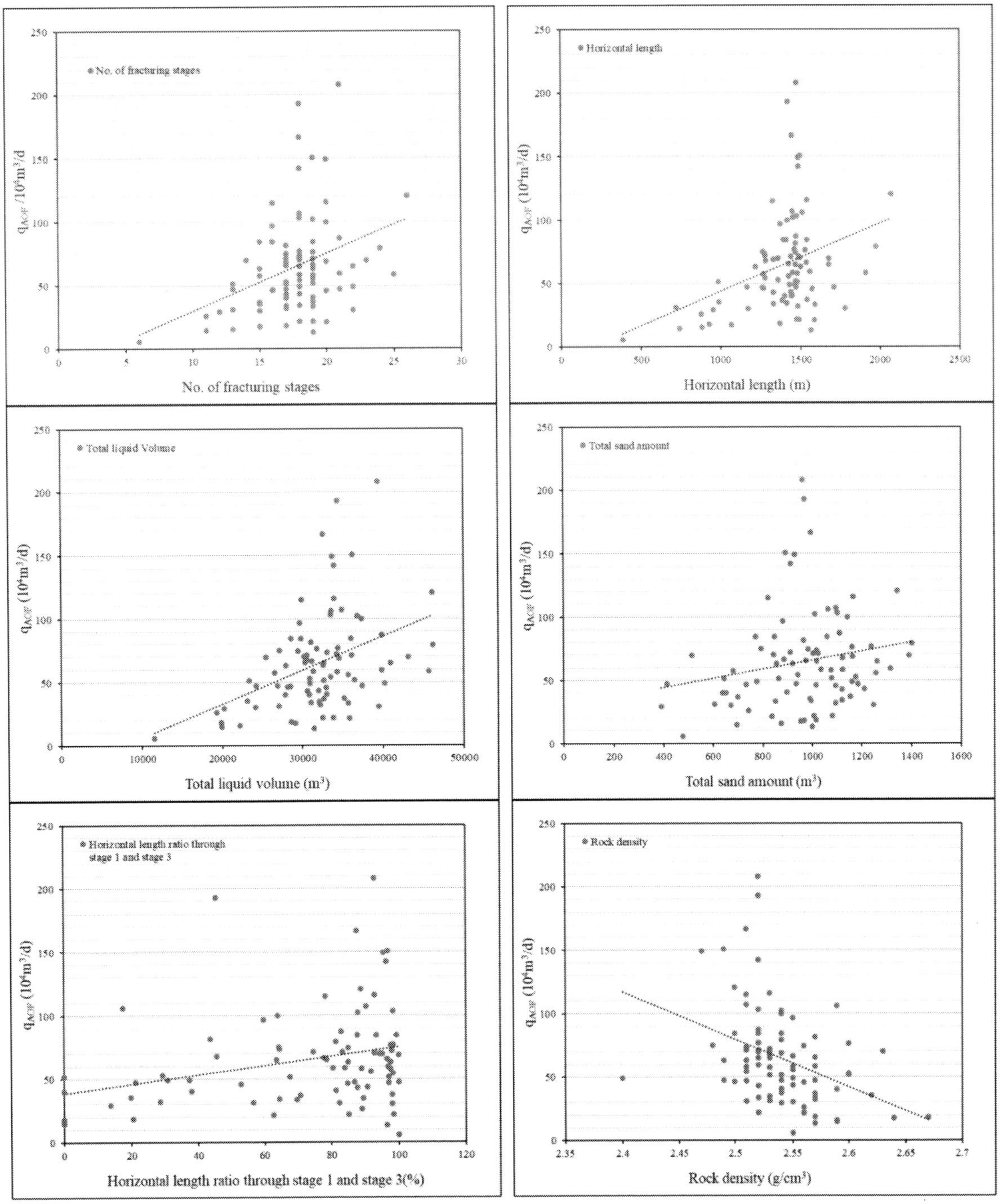
No. of fracturing stages
No. of fracturing stages
qAOF /10⁴m³/d
Horizontal length
Horizontal length (m)
qAOF (10⁴m³/d)
Total liquid Volume
Total liquid volume (m³)
qAOF (10⁴m³/d)
Total sand amount
Total sand amount (m³)
qAOF (10⁴m³/d)
Horizontal length ratio through stage 1 and stage 3
Horizontal length ratio through stage 1 and stage 3(%)
qAOF (10⁴m³/d)
Rock density
Rock density (g/cm³)
qAOF (10⁴m³/d)

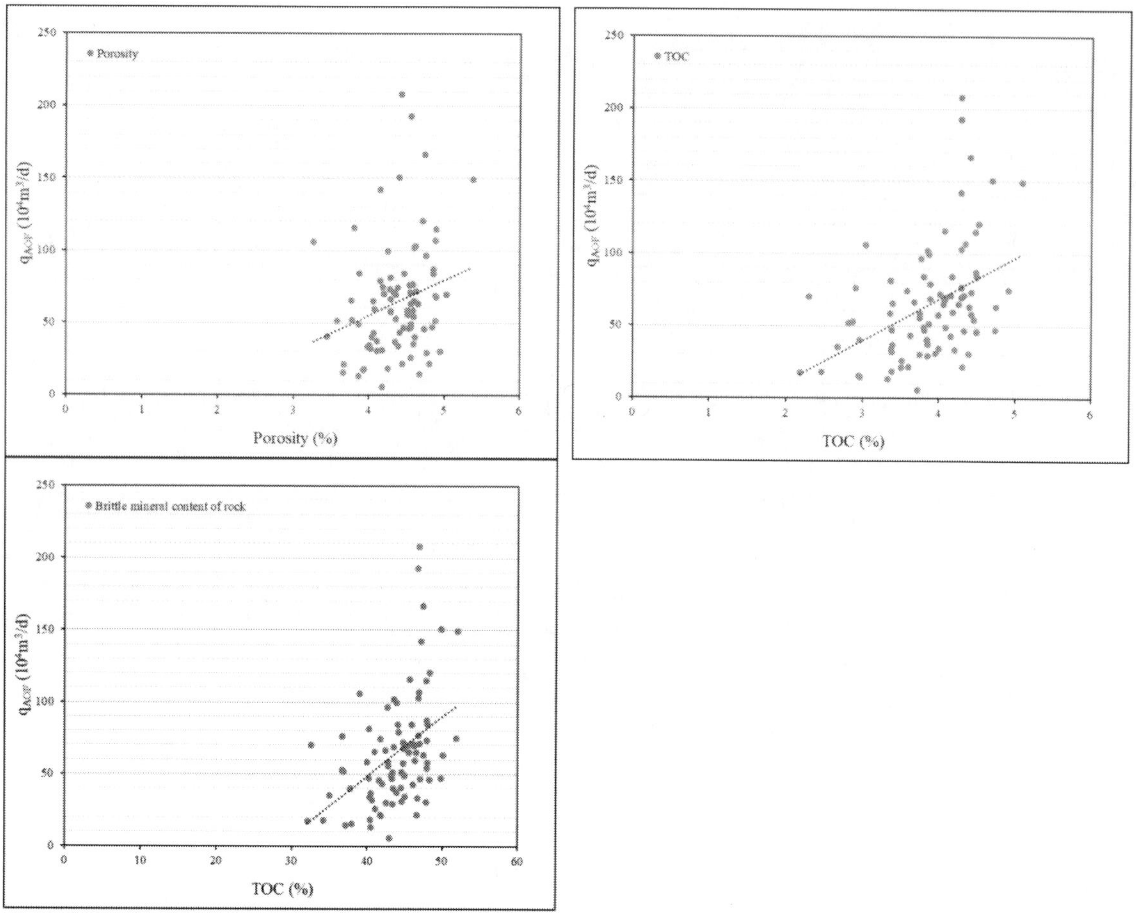

Fig.2 The scatter plots with single linear regression line of sample data of 86 gas wells in Jiaoshiba area with the AOF potential against 9 factors that may affect the deliverability of gas wells (i.e. number of fracture stages, horizontal length of well, total liquid volume, total sand amount, rock density, porosity, TOC, brittle mineral content of rocks, horizontal length ratio through stage 1 and stage 3 of Long1 member)

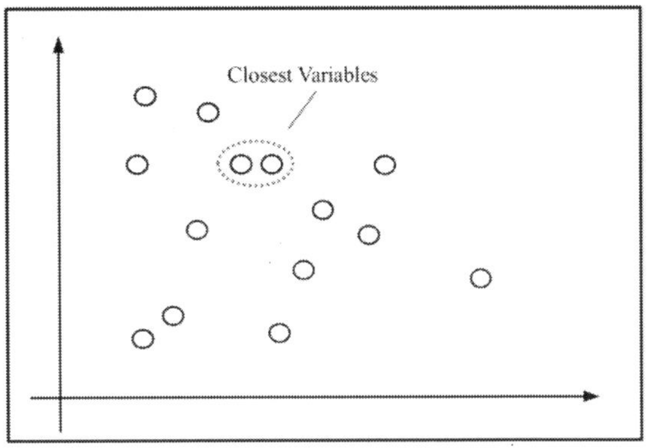

(a)

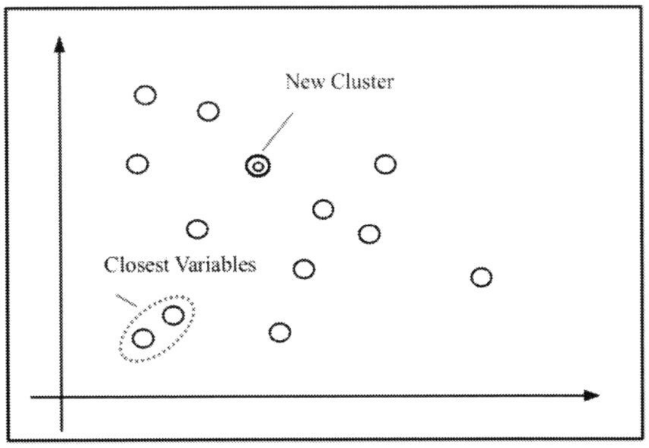

(b)

Fig.3 Hierarchical Clustering Analysis: (a) Clustering the variables by closest relation; (b) The center of new cluster is calculated by centroid method

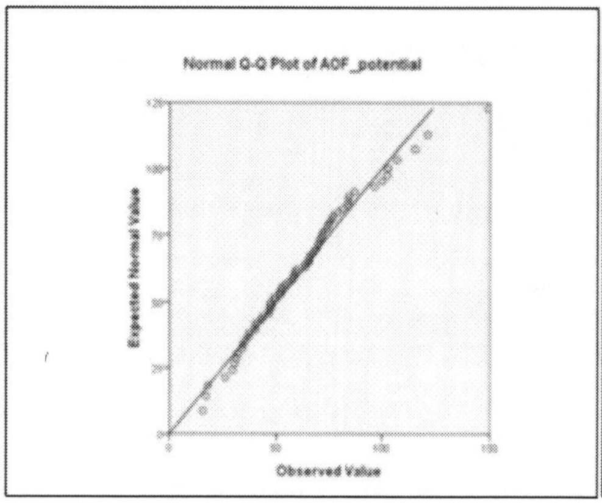
Normal Q-Q Plot of AOF_potential
Expected Normal Value
Observed Value

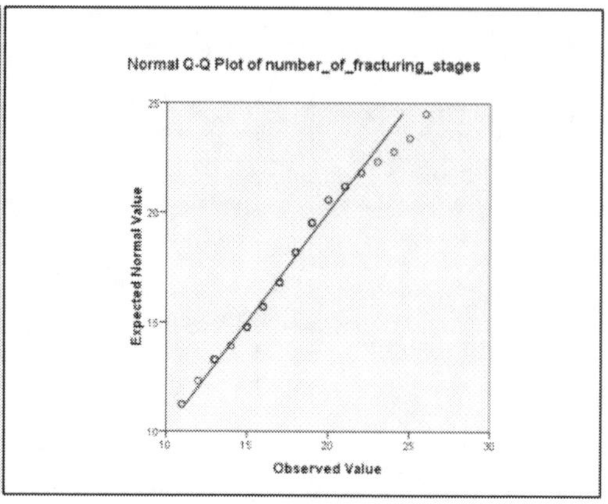
Normal Q-Q Plot of number_of_fracturing_stages
Expected Normal Value
25
20
15
10
10
15
20
25
30
Observed Value

a.

b.

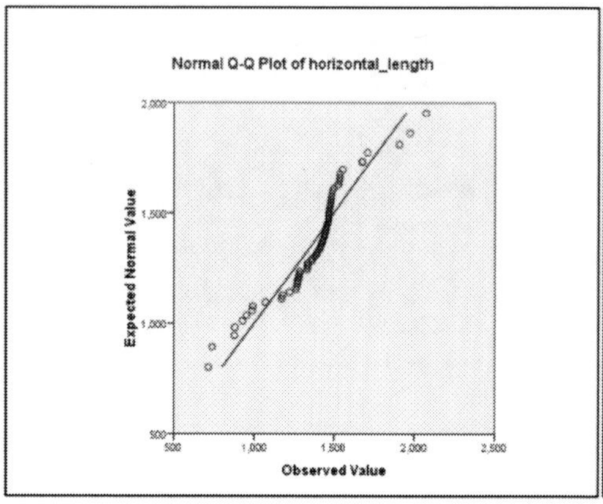
Normal Q-Q Plot of horizontal_length
Expected Normal Value
2,000
1,500
1,000
500
500
1,000
1,500
2,000
2,500
Observed Value

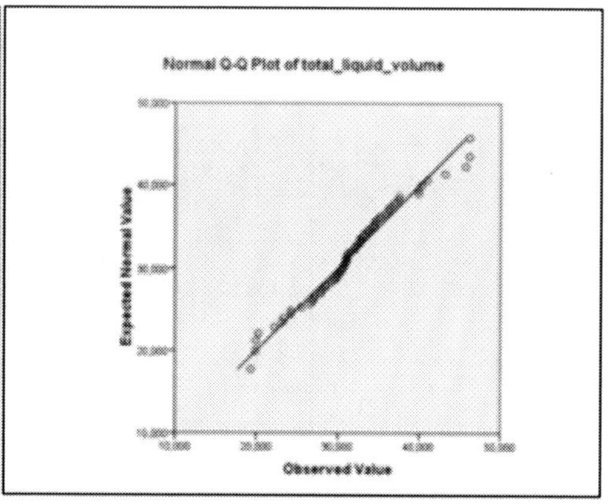
Normal Q-Q Plot of total_liquid_volume
Expected Normal Value
Observed Value

c.

d.

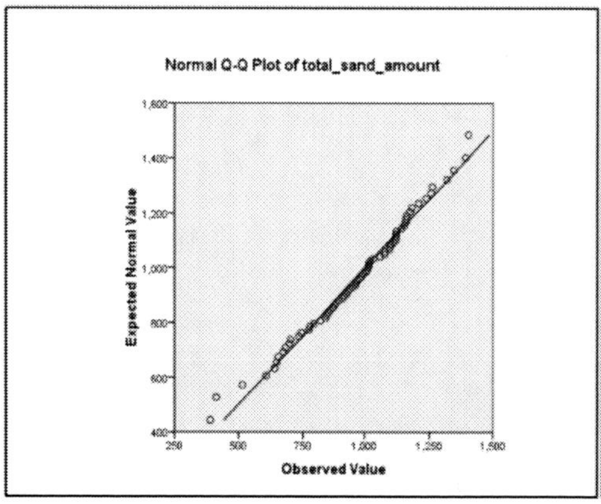
Normal Q-Q Plot of total_sand_amount
Expected Normal Value
1,600
1,400
1,200
1,000
800
600
400
250
500
750
1,000
1,250
1,500
Observed Value

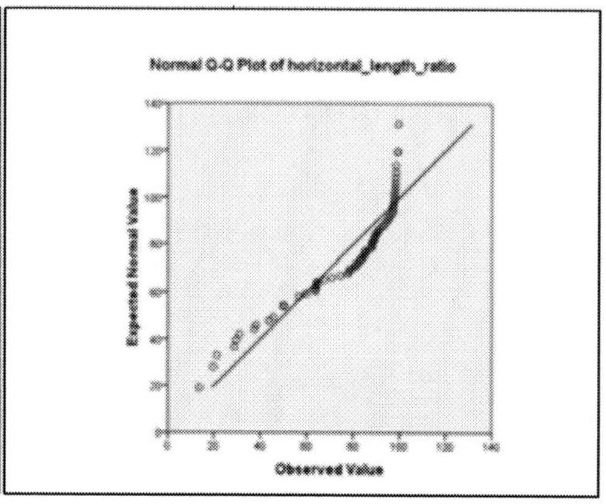
Normal Q-Q Plot of horizontal_length_ratio
Expected Normal Value
Observed Value

e.

f.

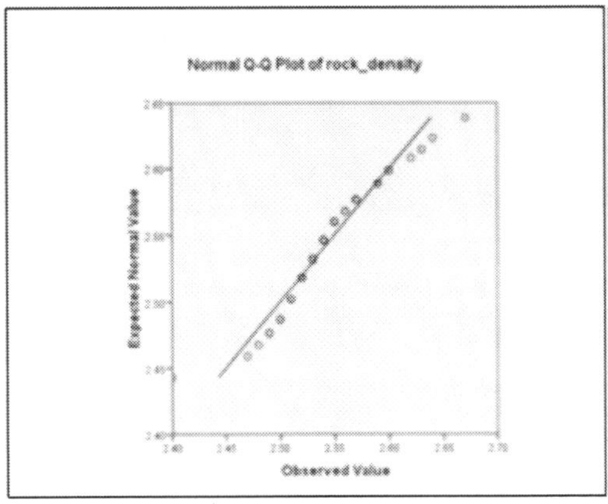

g.

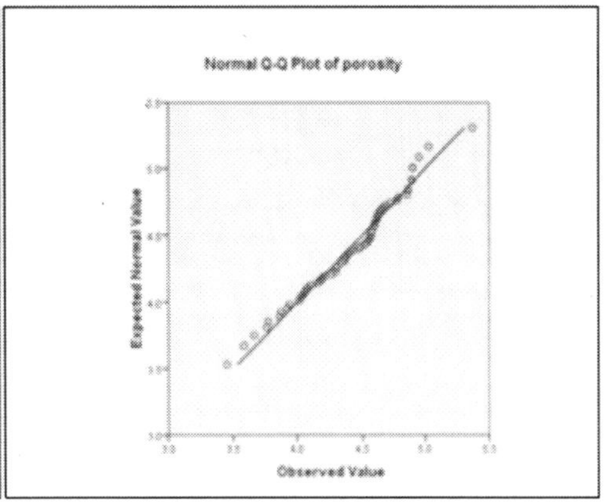

h.

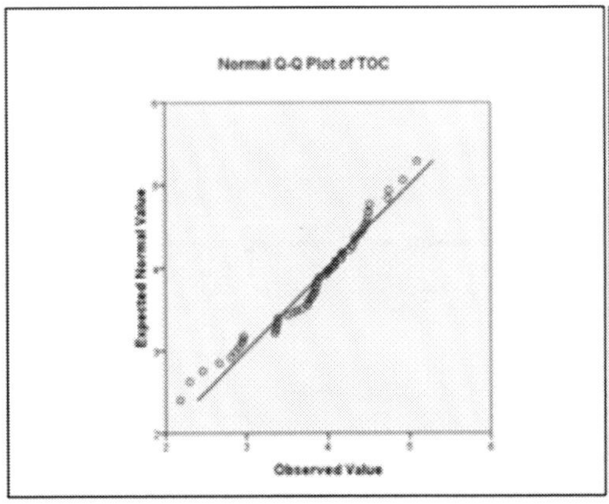

i.

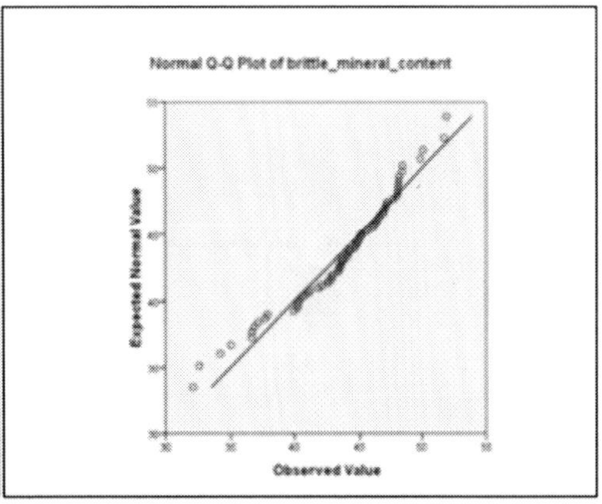

j.

Fig.4 The Q–Q (quantile-quantile) plots for the 10 variables of the 86 wells: a. AOF potential; b. number of fracturing stages; c. horizontal length of well; d. total liquid volume; e. total sand amount; f. horizontal length ratio through stage 1 and stage 3 of Long1 member; g. rock density; h. porosity; i. TOC; j. brittle mineral content of rock

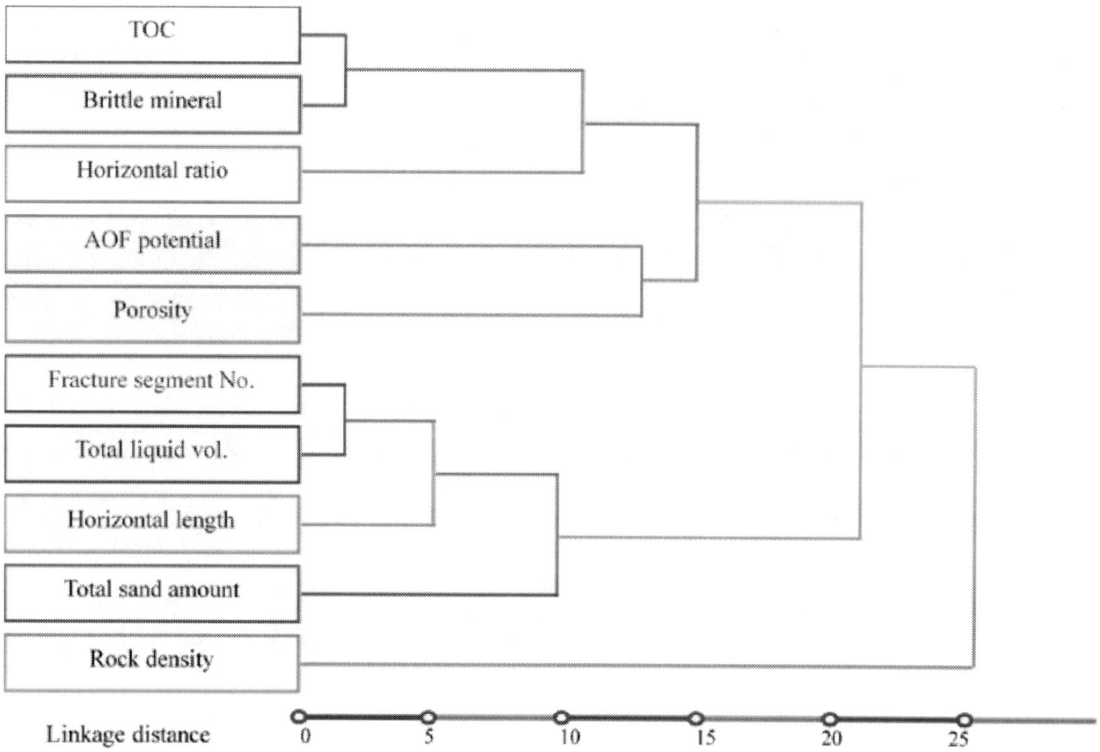

Fig.5 Hierarchical clustering analysis result of the sample data of 86 wells with linkage distance

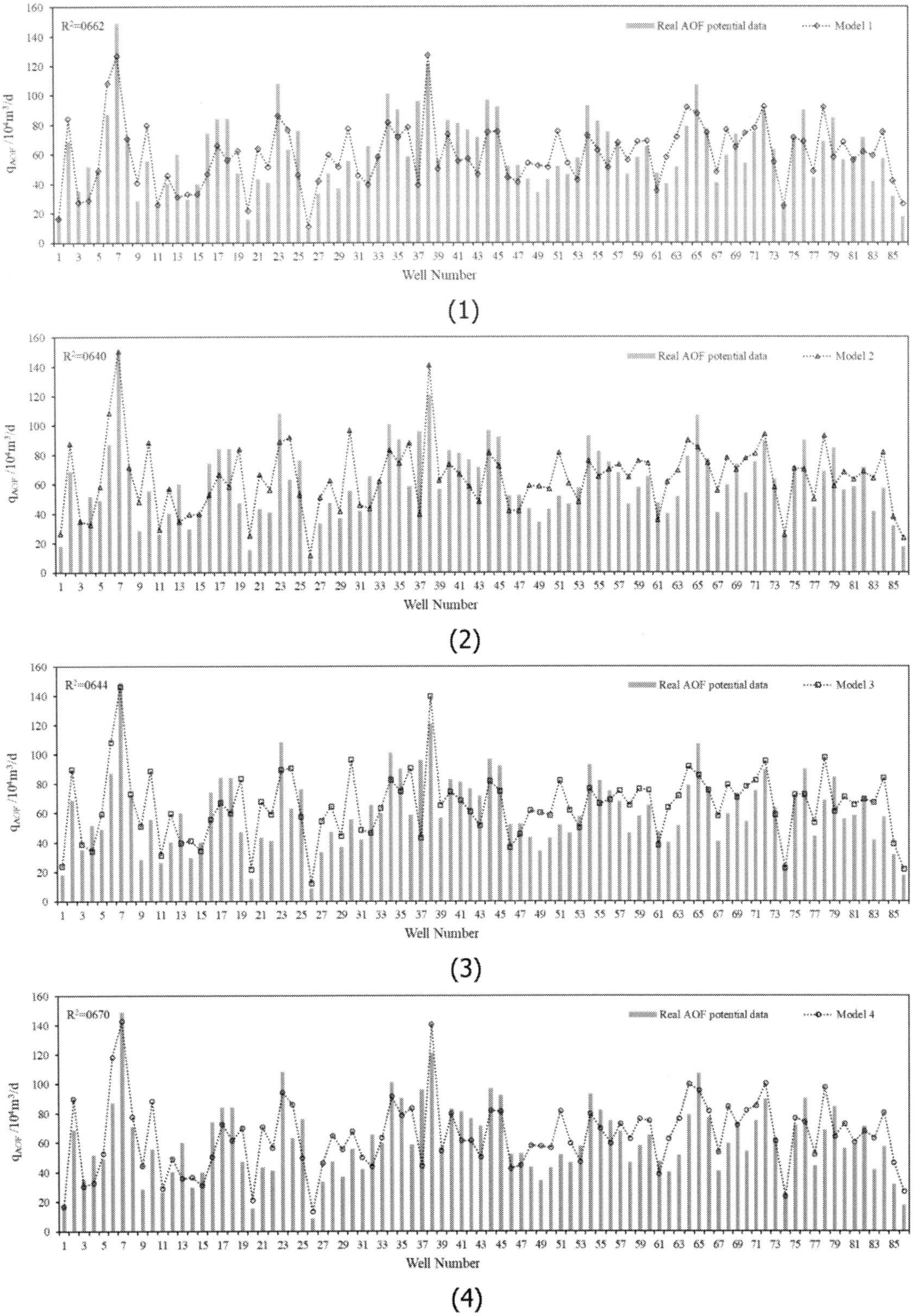

Fig.6 Comparison of actual and predicted AOF potential by the 4 models for the 86 wells with consideration of different key factors in Fuling Jiaoshiba shale gas field

Table 1 Correlation coefficients matrix of all variables in Jiaoshiba area

Factors	AOF potential	No. of fracturing stages	Horizontal length	Total liquid Volume	Total sand amount	Horizontal length ratio through stage 1 and stage 3	Rock density	Porosity	TOC	Brittle mineral content of rock
AOF potential	1	0.354	0.348	0.405	0.194	0.285	-0.373	0.246	0.433	0.438
No. of fracture stages	0.354	1	0.897	0.946	0.765	0.273	-0.203	-0.013	0.241	0.235
Horizontal length	0.348	0.897	1	0.863	0.633	0.331	-0.241	0.024	0.26	0.252
Total liquid Volume	0.405	0.946	0.863	1	0.736	0.302	-0.248	-0.01	0.274	0.282
Total sand amount	0.194	0.765	0.633	0.736	1	0.22	-0.002	-0.072	0.072	0.07
Ratio through stage 1 and stage 3	0.285	0.273	0.331	0.302	0.22	1	-0.492	0.287	0.568	0.578
Rock density	-0.373	-0.203	-0.241	-0.248	-0.002	-0.492	1	-0.289	-0.908	-0.917
Porosity	0.246	-0.013	0.024	-0.01	-0.072	0.287	-0.289	1	0.419	0.336
TOC	0.433	0.241	0.26	0.274	0.072	0.568	-0.908	0.419	1	0.982
Brittle mineral content of rock	0.438	0.235	0.252	0.282	0.07	0.578	-0.917	0.336	0.982	1

Table 2 Hierarchical clustering analysis result of factors in Jiaoshiba area

Parameter	Hierarchical clustering analysis result
AOF potential	1
Number of fracturing stages	2
Horizontal length	2
Total liquid volume	2
Total sand amount	2
Horizontal length ratio through stage 1 and stage 3	1
Rock density	3
Porosity	1
TOC	1
Brittle mineral content of rock	1

Table 3 Factors chosen in regression model according to hierarchical clustering analysis results

Regression model	Factors considered
1	a. Total liquid volume (X_1) b. Porosity (X_2) c. TOC (X_3)
2	a. Fracture segment No. (X_4) b. Porosity (X_2) c. TOC (X_3)
3	a. Fracture segment No. (X_4) b. Horizontal length ratio through stage 1 and stage 3 (X_5) c. TOC (X_3) d. Porosity (X_2)
4	a. Total liquid volume (X_1) b. Horizontal length ratio through stage 1 and stage 3 (X_5) c. Porosity (X_2) d. Brittle mineral content of rock (X_6)

601617 Solutions of C6F13- and C4F9-Based Polyfluoroacrylates in CO_2; Solubility, Viscosity, and Adsorption Onto Cement

Abstract:

High molecular weight poly(fluorinated acrylates) (PFAs) have drawn interest in the chemical and petroleum engineering fields due to their remarkable solubility in liquid and supercritical carbon dioxide (sCO_2) without the need for any co-solvent. This study does not include any examples of PFA based on monomers with the -C8F17 moiety because the ultimate hydrolysis degradation products of such polymers leads to the formation of PFOA, a compound deemed unsafe by the EPA. Instead, we have synthesized PFA based on fluoroacrylayte monomers containing the -C6F13 group (the functionality used in current commercial fluoroacrylate copolymers). These -C6F13-based polymers would degrade to form perfluorohexanoic acid (PFHxA) that, while significantly safer than PFOA, is also receiving more scrutiny from regulatory agencies in recent days. Therefore, we have also synthesized PFA based on monomers containing the -C4F9 functionality. The -C4F9-PFA ultimate degradation products include perfluorobutanoic acid (PFBA), which is significantly more environmentally benign than the -C6F13-based PFA. Therefore, one of the main objectives of this study was to determine if -C4F9-PFA retained the remarkable CO_2-solubility and CO_2-thickening capability as PFA based on the -C6F13 PFA and previously reported -C8F17-based PFA.

Many samples of PFA based on the -C6F13 and -C4F9-based fluoroacrylate monomers and varying ratios of initiator:monomer were bulk polymerized and then characterized for molecular weight and polydispersity. The -C4F9-PFA was shown to be nearly identical in CO_2-solubility as the -C6F16-PFA. Between 1-4 wt% PFA, the solubility of PFA in CO_2 was nearly independent of PFA molecular weight. Further, the cloud points loci for mixtures of CO_2-C4F9-PFA were nearly identical to the cloud point curves for the CO_2-C6F13-PFA at each temperature, with higher pressure required at higher temperature. This demonstrated that the -C4F9-PFA was just as CO2-soluble as the -C6F13-PFA. Further, both polymers were capable of thickening CO_2 in dilute concentration to a comparable degree (e.g. a 6-fold increase in viscosity with 4 wt% C4F9 PFA, 1.24×10^6 Da PFA in CO_2 at low shear rate).

PFA has previously been considered for use as a "CO_2 thickener" in enhanced oil recovery (EOR) operations. Unlike common chemical engineering applications where low viscosity is beneficial to reduce mass transfer resistance, low CO_2 viscosity is often problematic for EOR because it leads to viscous fingering (mobility control problems) and preferential loss of disproportionate amounts of CO_2 to high permeability "thief zones" that contain little recoverable oil (conformance control problems). However, in 2019 it was shown that PFA dissolved in CO_2 adsorbs so strongly onto sandstone or carbonate rock that the use of PFA in large-scale EOR projects to address mobility control would not be feasible, although near-wellbore conformance control applications (i.e. injecting the CO_2-PFA into isolated thief zones) may be viable.

In the current study, however, we are attempting to exploit the adsorption of PFA on solid surfaces for improved wellbore integrity. We are proposing to inject PFA-CO_2 solutions into small cracks within cement or between steel and cement (i.e. microannular cracks), thereby allowing a portion of the PFA to absorb and greatly reduce the crack permeability or completely seal the crack. Because PFA is a high molecular weight, amorphous, transparent, sticky, stretchy, oil-repellant, water-repellant polymer that is insoluble in water, oil or natural gas, it is a promising sealant for wellbore applications. Even though PFA is inherently expensive, this is a small volume application in which the cost of PFA would not exclude its consideration. Relative to other sealants (cement, resins, aqueous emulsions), PFA-CO_2 solutions are orders of magnitude less viscous, which would allow them deeper access into smaller cracks than other options. The PFA-CO_2 solutions would, however, be an inappropriate selection for the remediation of extremely large cracks or voids in cement.

In order to lay the groundwork for this proposed application, we have also determined the equilibrium adsorption of PFA onto Portland cement particles. This was accomplished by a direct measurement of the mass gain of a known quantity of sieved (22-30 mesh) Portland cement particles that was immersed in a PFA-CO2 fluid of specified original composition (e.g. 4 wt% PFA). The equilibrium concentration of PFA in the liquid CO_2 was determined via mass balance (original PFA placed in the cell – the amount of PFA found to adsorb onto the cement). This is, to the best of our knowledge, the first report of polymeric adsorption levels of solid surfaces immersed in CO_2-polymer solutions

611710 Competitive Sorption of CO2 in Shale Nanopores for Sequestration and Enhanced Gas Recovery Using Molecular Density Functional Theory

Abstract:

Enhanced gas recovery (EGR) by carbon dioxide (CO_2) injection to gas-bearing shale formations has been shown to be feasible and beneficial experimentally from pore scale to field scale. In addition to the improved recovery factor, CO_2 is trapped in the geological formation, which provides an opportunity to store CO_2. When supercritical CO_2 is injected, the preferential adsorption and dissolution of CO_2 in nanopores and organic matter enables the displacement of shale gas. This mechanism has been investigated by molecular simulation in graphite pores, but rarely for permeable pores by simulation or classical density functional theory (DFT). The compositional distribution of components from the organic matter source (kerogen) to nanoscale pores in the organic matter and to large fractures makes the estimation of fluid storage and the multicomponent compositional distributions in shale a challenging problem. We have introduced a density functional theory that accounts for molecular size and shape and kerogen maturity and swelling to predict the partitioning of components between the kerogen matrix and nano-pores in the kerogen [1].

In this work, a molecular thermodynamic model is presented that predicts both the adsorption of hydrocarbons in nanoscale pores and the dissolution of fluids in organic matter in equilibrium with the bulk kerogen phase. The swelling ratios of five kerogens of varying maturities and types in different solvents are well quantified using the new model. This kerogen matrix model is incorporated in interfacial statistical associating fluid theory (iSAFT) DFT to form a nanoporous kerogen composite model. The equilibrium partitioning of mixtures of fluids from the bulk phase to the nanoporous kerogen phase show enrichment of aromatic and cyclic molecules in kerogen, the preferential adsorption/absorption of high molecular weight molecules and carbon dioxide, and the significant amount of hydrocarbon storage both in pore space and kerogen matrix [1, 2].

Further, we apply molecular DFT to study the competitive sorption of CO_2 with shale gas in kerogen at various conditions. The DFT model is verified by grand canonical Monte Carlo (GCMC) simulation in graphite slit pores for pure and binary component systems at different temperatures, pressures, pore sizes and bulk compositions for methane/ethane/ CO_2. [2]. The model is utilized to predict the CO_2 adsorption selectivity in multicomponent systems which include two different shale gases. We show the selectivity of CO_2 decreases as temperature / pressure/ nanopore size/ shale gas average molecular weight increase. Subsequently, we model the CO_2/CH_4 adsorption in pores of more realistic conditions, in which we consider different levels of moisture in the pore and the gas dissolution in organic matrix. Interestingly, the presence of water inside a nanopore greatly reduces the CO2 adsorption but improves the CO_2 selectivity with respect to methane. The presence of organic matter kerogen also increases the CO_2 selectivity. In summary, a novel permeable pore DFT model enables prediction of partitioning and distribution of CO2 and hydrocarbons in shale related to enhanced gas recovery and CO_2 storage. [1, 2].

1. Jinlu Liu and Walter G. Chapman, "Thermodynamic Modeling of the Equilibrium Partitioning of Hydrocarbons in Nanoporous Kerogen Particles," Energy & Fuels, 33, 891-904 (2019).
2. Jinlu Liu, Shun Xi, and Walter G. Chapman, "Competitive Sorption of CO2 with Gas Mixtur in Nanoporous Shale for Enhanced Gas Recovery from Density Functional Theory," Langmuir, 35, 8144-8158 (2019).

609261 Spatial and Temporal Evolution of Natural Gas and Natural Gas Liquid Feedstocks in Shale Basins

Abstract:

Technological advancements in horizontal drilling and hydraulic fracturing have led to increasing production of light hydrocarbons from shale resources, including natural gas and natural gas liquids (NGLs). However, some basins that have insufficient gas gathering and pipeline networks (e.g. the Permian Basin in Texas) may flare significant quantities of gas. For example, in the Permian Basin, about 4.6% of gross gas production was estimated to be flared between May 2018 and March 2019 (Zhang et al., 2020). To mitigate flaring issues, hydrocarbon resources can be locally transformed to chemical or energy products. The National Science Foundation's Center for Innovative and Strategic Transformation of Alkane Resources (CISTAR) is developing modular processes to transform natural gas and NGLs to gasoline-like transportation fuels.

Determining the scale of operation of such modular chemical manufacturing requires a comprehensive understanding of the temporal and spatial availability of the hydrocarbon feedstocks. Preliminary work was performed in the Eagle Ford oil and gas producing region in south central Texas. With over 20,000 wells, the Eagle Ford produced 5.4 billion cubic feet natural gas per day and 1.2 million barrels of condensate and oil per day in 2019 (RRC, 2020). From the northwest to the southeast portion in the Eagle Ford Shale, well production varies from oil, to gas with condensate (wet gas), and to dry gas, as shown in Figure 1 (Gherabati et al, 2016). From oil to wet gas to dry gas regions, propane to ethane ratio varies from 1:1 to 1:3 (mass basis). In addition to varying gas compositions, the magnitude of available feedstocks also vary across the region. Ethane available at individual gas plants was estimated by assuming that produced gas from wells is sent to their nearest gas plant. Among 44 gas processing plants in the Eagle Ford, 60% of ethane production are represented by 43 plants with 200,000 – 750,000 kg of ethane per day, with one plant processing at ~3,000,000 kg/day, accounting for 40% of ethane production, shown in Figure 2.

Similar work has been performed in the Permian Basin in Texas. The Permian Basin is one of the largest oil-producing regions in the world and accounts for over 30% of oil production in the United States. Recent study (Zhang et al, 2020) shows that methane emissions in the Permian Basin are the highest among all the U.S. oil/gas producing regions. This presentation will provide detailed characterizations of spatial and temporal variability in light alkane production, and losses due to flaring for a case study region in the Permian Basin.

Collectively, the analyses in the Eagle Ford and the Permian Basin will provide two case studies of spatial and temporal characteristics of feedstock availability for modular light alkane manufacturing.

References

Gherabati, S. A., Browning, J., Male, F., Ikonnikova, S. A., & McDaid, G. (2016). The impact of pressure and fluid property variation on well performance of liquid-rich Eagle Ford shale. *Journal of Natural Gas Science and Engineering*, *33*, 1056-1068.

Railroad Commission of Texas (RRC) – Eagle Ford Shale Information. (2020). Available at: https://www.rrc.state.tx.us/oil-gas/major-oil-and-gas-formations/eagle-ford-shale-information/

Zhang, Y., Gautam, R., Pandey, S., Omara, M., Maasakkers, J. D., Sadavarte, P., ... & Zhang, R.

(2020). Quantifying methane emissions from the largest oil-producing basin in the United States from space. *Science Advances*, *6*(17), eaaz5120.

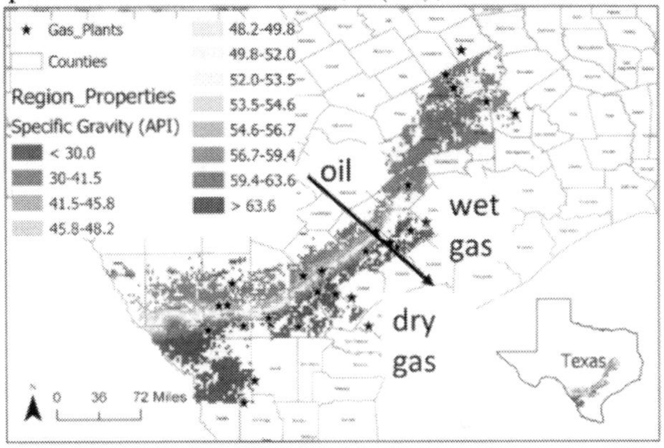

Figure 1. Spatial variabilities in well production in the Eagle Ford Shale.

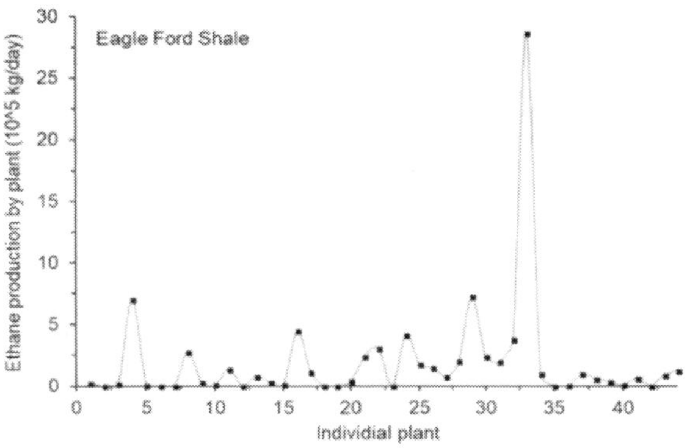

Figure 2. Ethane available at 44 gas processing plants in the Eagle Ford Shale

592706 A New Geology-Engineering Integration Solution for Shale Gas Exploration: A Case Study from Changning Block in China

Abstract:

Rich and quality shale gas reservoirs have been found in the Lower Silurian Longmaxi strata in the southern Sichuan Basin, which was identified as great development potential. However, shale gas reservoir is characterized by ultra-low permeability, which usually has no natural productivity and is called a typical "artificial gas reservoir". We must pay attention to reservoir quality, drilling quality and completion quality in order to obtain the maximum EUR and the highest oil recovery. To tackle this challenge, the integrated Geoengineering solution was deployed in Changning Shale Gas Block, which consists of geological modeling, geomechanics, fracturing simulation and gas reservoir simulation. Geoengineering integration technology is used on the Changning A platform for field applications. The results show that: (1) Carrying out integrated research on geological engineering can realize the organic combination of geology and engineering. The geological model provides technical support for engineering optimization. The engineering practice constantly corrects the geological model and improves the precision; (2) Through the integration of geological engineering, we can create "transparent" shale gas reservoir, which can improve the predictive accuracy of geological and engineering sweet spot mapping; (3) Hydraulic fractures can be fine characterization by using geological engineering integration technology, which is useful for quantitative description and analysis of key parameters such as geometrical shape, net pressure, proppant distribution and conductivity of hydraulic fractures; (4) the refined unstructured-grid (<10m) based 3D numerical simulation model was built for hydraulic fractured horizontal wells in shale gas reservoir, the history match for shale gas wells has been hugely improved: gas production rate was matched to more than 98%; the wellhead pressure was matched to more than 80%; the cumulative liquid production rate was matched to more than 90%. With the implementation of the Geoengineering study, the shale gas production in Changning area has been increasing rapidly in the past 2 years.

593227 New Technology of EOR of Ultra Deep and Complex Condensate Gas Reservoirs in Tarim Basin， China

Abstract:

A number of condensate gas reservoirs (both sandstone and carbonate) with depth of over 5000m have been developed in Tarim Basin, China. Those reservoirs are featured with thick reservoirs pay zones (over 100 meters) with ,strong reservoir heterogeneities in both rock properties and fluid properties. Gas channeling in the process of gas injection development and retrograde condensation in the process of depletion development have seriously affected the recovery of condensate. In order to enhance the condensate recovery in those condensate gas reservoirs, a series of new technologies have been developed during the last decade.

Laboratory experiment analysis and development practice discovered that three major mechanisms controlling the gas migration process in the reservoirs: displacement, gravity and diffusion for cyclic gas injection. It is proposed that gravity differentiation is the main factor affecting the development effect of cyclic gas injection, and the fluid distribution law of underground condensate gas reservoir is re-recognized. For the sandstone condensate gas reservoir developed with large well spacing (over 1000 meters), the displacement and diffusion will be replaced by gravity differentiation gradually in the process of cyclic gas injection, and the fluid will be redistributed vertically according to its density, which makes different phase behaviors. Similarly, the reservoir space of depleted carbonate condensate gas reservoir is mainly dissolution cave and fracture. Under the action of gravity, the reverse condensate oil usually accumulates at the bottom of fracture cave system. On this basis of this understanding, the technology of gravity assisted three-dimensional displacement gas injection for sandstone condensate gas reservoirs and horizontal well placement technology for carbonate condensate gas reservoirs were developed to optimize the well placement. The technologies have been successfully applied to the developed condensate gas field in Tarim Basin, and the recovery rate of condensate oil has been increased by over 19%.

Production capacity evaluation method of condensate gas reservoir considering the influence of stress sensitivity and condensate oil

DU Chunxiao, LIU Xiaohong, ZHANG Fengyi, XU Jing, GENG Zhigang

(Tianjin Branch of CNOOC Ltd., Tianjin 300459)

Abstract: Accurate evaluation of production capacity is an important guarantee for efficient development of condensate gas reservoir. For the deep condensate gas reservoir with low permeability and low pressure difference between the initial formation pressure and the dew point, the influence of reservoir stress sensitivity and condensate oil on the production capacity of the gas reservoir cannot be ignored. Nevertheless, at present, the widely used method to calculate the gas reservoir production capacity is based on single-phase gas, without considering the influence of condensate oil and stress sensitivity on reservoir permeability. On the basis of researching a large number of literatures, combined with laboratory experiments and numerical simulation, the production capacity evaluation model of condensate gas reservoir under the comprehensive influence of stress sensitivity and condensate oil is derived by using the seepage flow mechanics theory, and the research results are applied to the Bohai Z gas reservoir. The results show that the production capacity evaluation model can more accurately reflect the production status of this kind of gas reservoir in different development stages, and can be widely used in the productivity evaluation of similar condensate gas reservoir.

Key words: deep condensate gas reservoir; stress sensitivity; condensate oil; evaluation of production capacity; low permeability

Gas reservoir production capacity evaluation is an important research content of gas reservoir engineering research, which provides key guidance for the design of gas reservoir development plan. For condensate gas reservoir, the fluid flow law and phase change are more complex, and the reverse condensation phenomenon has a serious impact on gas well productivity. According to theoretical and experimental studies, many scholars have confirmed that low permeability reservoirs have strong stress sensitivity. At present, the widely used method to calculate the gas reservoir production capacity is based on single-phase gas, without considering the influence of condensate oil and stress sensitivity on reservoir permeability. Many scholars have carried out research on the influence of reservoir stress sensitivity or retrograde condensation on gas well productivity, but lack of comprehensive consideration of both. Therefore, a new productivity evaluation method of condensate gas reservoir considering stress sensitivity and retrograde condensation is derived by combining laboratory experiments and numerical simulation methods, and is applied to BZ gas reservoir in Bohai oilfield.

1. Influence of stress sensitivity on reservoir permeability

The relationship between rock permeability and effective confining pressure has been studied deeply by many scholars using a large number of experiments. The relationship between permeability and effective confining pressure can be divided into three types: power-law relationship, exponential relationship and polynomial relationship.

BZ gas reservoir in Bohai oilfield is a low permeability gas reservoir. The core stress sensitivity test results of BZ-1 well are shown in Table 1 and figure 1. It can be seen that the dimensionless permeability of core decreases with the increase of effective stress, and BZ gas reservoir has strong stress sensitivity.

Table 1 Data of core stress sensitivity test

Net pressure MPa	Sample 1 K_n/K_i	Sample 2 K_n/K_i	Sample 3 K_n/K_i	Sample 4 K_n/K_i
3.5	0.748	0.792	0.774	0.781
5.0	0.637	0.670	0.642	0.643
7.0	0.476	0.521	0.503	0.524
9.0	0.391	0.394	0.402	0.414
11.0	0.313	0.329	0.308	0.369
15.0	0.208	0.238	0.239	0.248
20.0	0.152	0.170	0.180	0.170
25.0	0.117	0.121	0.141	0.131
30.0	0.095	0.102	0.122	0.112
35.0	0.081	0.087	0.097	0.083

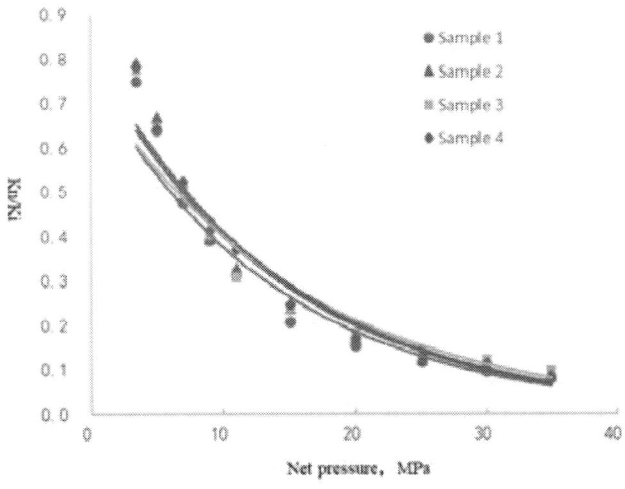

Fig 1 Stress sensitivity curves of the cores

By fitting the relationship between rock permeability and effective confining pressure, it is found that there is a good exponential relationship between them（Table 2）. It can be expressed as:

$$k_n = \mathrm{N}k_i e^{-\alpha(p_i - p)} \tag{1}$$

Table 2 Stress sensitivity index fitting of core permeability

Core number	Exponential fitting expression	R^2
1	$y = 0.7721e^{-0.072x}$	0.9553

2	$y = 0.8186e^{-0.071x}$	0.9595
3	$y = 0.7609e^{-0.064x}$	0.9489
4	$y = 0.8331e^{-0.07x}$	0.9727

According to the principle of seepage mechanics, the square distribution expression of stable seepage pressure of gas reservoir is as follows.

$$p^2 = {p_e}^2 - \frac{{p_e}^2 - {p_{wf}}^2}{\ln(\frac{r_e}{r_w})} ln\frac{r_e}{r} \tag{2}$$

Formula (1) is deformed.

$$k_n = Nk_i e^{-\alpha\left\{p_i - \left[{p_e}^2 - \frac{{p_e}^2 - {p_{wf}}^2}{ln(\frac{r_e}{r_w})} ln\frac{r_e}{r}\right]^{\frac{1}{2}}\right\}} \tag{3}$$

2 Influence of condensate on gas phase relative permeability

The influence of reverse condensate oil is studied by numerical simulation method. By using single well numerical simulation, the variation law of condensate oil saturation at different distances from well during gas field development is obtained.

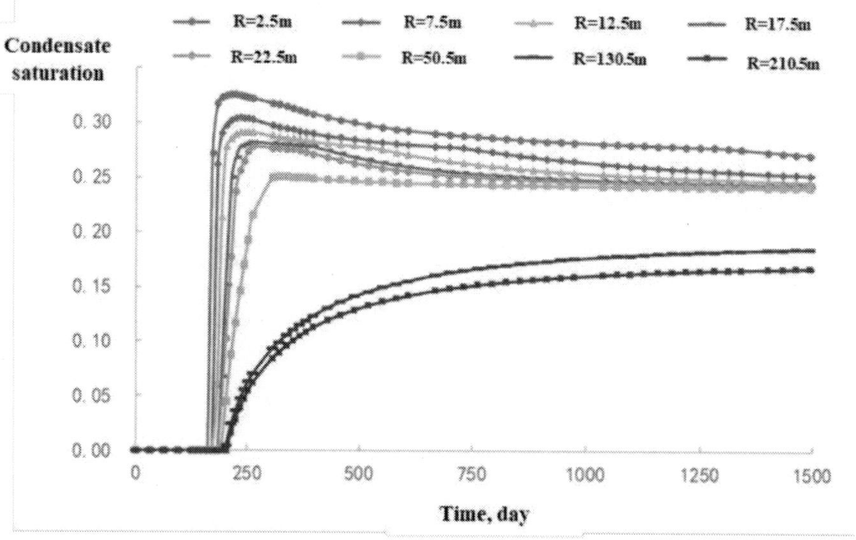

Fig.3 variation law of condensate oil saturation

It can be seen from Figure 3 that the condensate saturation decreases with the increase of radial distance, and tends to be stable after increasing with the extension of production time. The results show that the condensate saturation changes with time in the near well zone, and the distribution function of condensate saturation with radial distance $s_o(\mathrm{r})$ can be obtained at any time.

According to the relative permeability curve, the regression relationship between gas relative permeability and oil saturation can be obtained. The relative permeability curve of BZ gas reservoir in Bohai oilfield is shown in Fig. 4, the regression expression is:

$$k_{rg}(\mathrm{r}) = 4.7367s_o(r)^3 - 2.0497s_o(r)^2 - 2.1398s_o(r) + 1.0015 \tag{4}$$

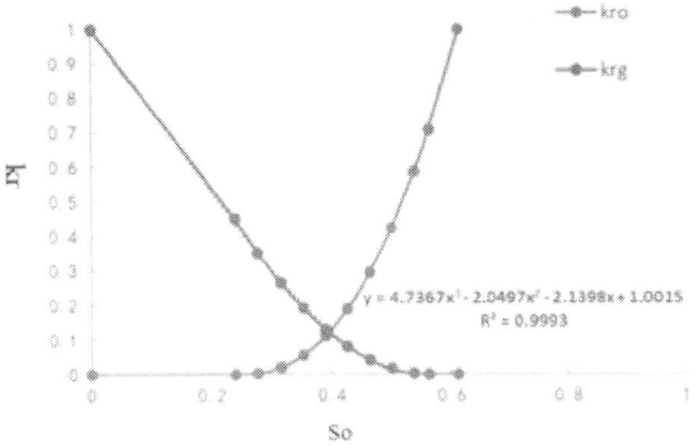

Fig.4 Relative permeability curve of Z gas reservoir

3 Improved gas well productivity evaluation model

The expression of effective permeability is:

$$k_g = k_{\mathrm{n}} \cdot k_{rg} \tag{5}$$

Replace formula (3) and formula (4) into formula (5):

$$k_g = Nk_i[4.7367s_o(r)^3 - 2.0497s_o(r)^2 - 2.1398s_o(r) + 1.0015]e^{-\alpha\{p_i - [p_e{}^2 - \frac{p_e{}^2 - p_{wf}{}^2}{\ln(\frac{r_e}{r_w})} ln\frac{r_e}{r}]^{\frac{1}{2}}\}} \tag{6}$$

The condensate saturation beyond a certain distance from the well is lower than the critical flow saturation, the distance is defined as the effective radial distance and expressed as r_o.

The whole stratum can be divided into two regions. The effective permeability can be expressed as:

$$k_g(\mathrm{r}, p_{wf}) = \begin{cases} Nk_a[4.7367s_o(r)^3 - 2.0497s_o(r)^2 - 2.1398s_o(r) + 1.0015]e^{-\alpha\{p_i - [p_e{}^2 - \frac{p_e{}^2 - p_{wf}{}^2}{\ln(\frac{r_e}{r_w})} ln\frac{r_e}{r}]^{\frac{1}{2}}\}}, & r_w < r < r_o \\ Nk_a e^{-\alpha\{p_i - [p_e{}^2 - \frac{p_e{}^2 - p_{wf}{}^2}{\ln(\frac{r_e}{r_w})} ln\frac{r_e}{r}]^{\frac{1}{2}}\}}, & r_o \le r \le r_e \end{cases} \tag{7}$$

① $r_w < r < r_o$, average permeability can be expressed as:

$$k_1(p_{wf}) = \frac{\int_{r_w}^{r_o} k_g(r, p_{wf}) \cdot 2\pi r dr}{\int_{r_w}^{r_o} 2\pi r dr} \tag{8}$$

② $r_o \le r \le r_e$, average permeability can be expressed as:

$$k_2(p_{wf}) = \frac{\int_{r_o}^{r_e} k_g(r, p_{wf}) \cdot 2\pi r dr}{\int_{r_o}^{r_e} 2\pi r dr} \tag{9}$$

According to the principle of seepage mechanics, the average gas permeability of the formation can be obtained:

$$k(p_{wf}) = \frac{k_1(p_{wf}) \cdot k_2(p_{wf}) \cdot \ln(\frac{r_e}{r_w})}{k_2(p_{wf}) \cdot \ln(\frac{r_o}{r_w}) + k_1(p_{wf}) \cdot \ln(\frac{r_e}{r_o})} \tag{10}$$

Based on the above research, the productivity evaluation formula of gas well considering stress sensitivity and condensate oil is obtained.

$$p_e^2 - p_{wf}^2 = \frac{1.273\times10^{-3}T\mu Z}{k(p_{wf})h}\ln\left(\frac{r_e}{r_w}\right) q_{sc} + \frac{2.162\times10^{-10}\gamma_g ZT}{r_w h^2 k(p_{wf})^{1.5}} q_{sc}^{\ 2} \qquad (11)$$

4 Application

BZ is a rare condensate gas reservoir with complex reservoir and fluid characteristics. The main performance is low pressure difference between the initial formation pressure and the dew point (0.21-7.13MPa), the maximum amount of condensate is 40%, the reservoir permeability is 1.00~6.80mD, strong stress sensitivity.

The evaluation equation is applied to BZ gas field. The IPR curves of three cases are calculated, under the comprehensive influence of condensate oil and stress sensitivity, only considering the influence of condensate oil or stress sensitivity. The results are compared with those of the traditional binomial formula.

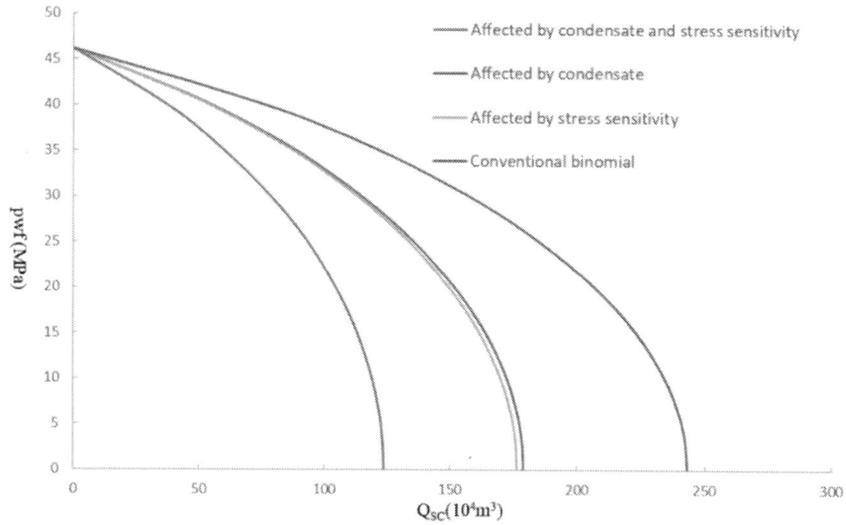

It can be seen that reservoir stress sensitivity and retrograde condensation have great influence on gas reservoir productivity. Under the combined effect of both, the productivity of gas reservoir is reduced by 50%. The IPR curves of gas wells under different formation pressures under the combined influence of stress sensitivity and retrograde condensation are further obtained.

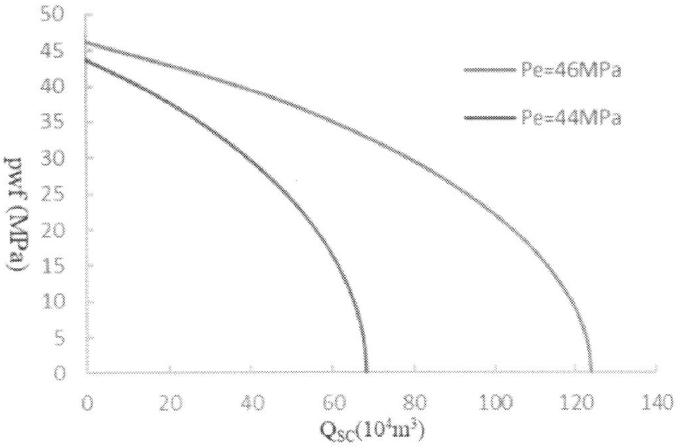

It can be seen that the formation pressure decreases, stress sensitivity and retrograde condensation increase, the productivity of gas wells also decreases obviously. The formation pressure drops by 2MPa, and the open flow rate of gas wells decreases by 57%. Therefore, in order to ensure the gas well productivity, it is necessary to adopt the development mode of gas injection and pressure maintaining, and the gas well production allocation should not be too high, in order to slow down the adverse impact of pressure drop.

6 Conclusion

(1) Stress sensitivity and condensate oil have great influence on the productivity of low permeability condensate gas reservoirs with low pressure difference between the initial formation pressure and the dew point. Most studies lack of comprehensive consideration of the two factors.

(2) Combined with laboratory experiment and numerical simulation, the formation permeability considering the effect of reservoir stress sensitivity and retrograde condensation is derived by using the principle of seepage mechanics, and then the evaluation method which can reflect the real productivity of gas reservoir is obtained.

(3) The research results show the necessity of gas injection to maintain reservoir pressure, and put forward suggestions for reasonable production of gas wells.

Symbol description

k_i-Original permeability, $10^{-3}\mu m^2$

k_n-Absolute permeability, $10^{-3}\mu m^2$

k_g-Effective permeability, $10^{-3}\mu m^2$

k_{rg}-Gas phase relative permeability, $10^{-3}\mu m^2$

p_i-Original formation pressure, MPa

p-Formation pressure, MPa

N-Stress sensitivity coefficient

α-Stress sensitivity index

p_e-Formation pressure at supply radius, MPa

p_{wf}-Bottom hole pressure, MPa

r_e-Supply radius, m

r_w-Wellbore radius, m

s_o-Condensate saturation near wellbore

r-Radial distance from well, m

T-Formation temperature, K

μ-Gas viscosity, mPa·s

Z-Gas deviation factor

h-Thickness of gas reservoir, m

β-Characteristic parameters of pore structure affecting turbulence and inertia resistance

γ_g-Relative density of natural gas

q_{sc}-Gas well production under standard conditions, m^3/d

References

[1] Fussel D.D. Single well Performance for gas condensate reservoir[J]. JPT，1973，25(12)：860-876.

[2] Jones J.R. and Raghavan R. Interpretation of Pressure buildup reservoirs in gas condensate well[J]. SPEFE，1989.

[3] FAN Xueping, XU Xiangrong. Experimental and mechanism research about permeability damage with the change of stress[J]. PETROLEUM EXPLORATION AND DEVELOPMENT, 2002

[4] XUE Zhan, LIU Pengcheng. A New Method of Correction of Gas Condensate Well Productivity Considering Capillary Number[J]. Journal of Chongqing University of science and Technology (NATURAL SCIENCE EDITION), 2014.

[5] LUO Ruilan, CHENG Linsong. A new method of determining relationship between permeability and effective overburden pressure for low-permeability reservoirs[J]. Journal of China University of Petroleum, 2007.

Numerical simulation study of a complex naturally-fractured gas reservoir with extra-high condensate in Bohai Oilfield

GAO Zhennan, SU Yanchun, GENG Zhigang, LIU Weilin, HUANG Baogang

(CNOOC China Limited Tianjin Branch, Tianjin 300459, China)

Abstract:

The BZ condensate gas field is the largest natural gas field in the history of the Bohai bay basin, with proven reserves of more than 100 billion cubic meters of natural gas and 100 million cubic meters of condensate oil. This gas field has unique features, such as deep burial, fracture development, strong stress sensitivity, small pressure difference between formation pressure and dew point pressure, high condensate content, etc. There is no similar condensate-gas field development cases in the world and proper reservoir numerical simulation study is crucial for efficiently developing such a reservoir.

This paper develops a numerical simulation flowchart for this kind of complex naturally fractured condensate gas reservoirs. First, based on logging, core and well testing data, the reservoir of BZ gas field is divided into three different types of deposits in the vertical direction: glutenite, semi-weathered zone and inner buried hill. The fluid flow among the three different types of deposits is calculated based on the equivalent fluid flow conduction between adjacent grids. Secondly, the correlation of permeability changes with stresses is established based on experimental data and implemented in the reservoir simulation based on the imbedded programming module in the simulator. Thirdly, through PVTi numerical simulation, EOS are tuned based on 7 pseudo-components. Finally, a dual-porosity compositional model is established, and the numerical simulation study is carried out to optimize the development plan. The developed flowchart will be a very useful tool for modeling naturally-fractured condensate gas reservoirs.

The simulation results indicated that: 1) considering the strong stress sensitivity, gas injection can effectively increase the permeability of the reservoir and control the condensate precipitation; 2) considering the structural characteristics of the reservoir, gas injection in the high position can form top-gas drive and expand the swept volume; 3) considering the influence of fracture, injection- production line perpendicular to fracture trend can effectively delay the breakthrough of injected gas and expand the injection capacity; 4) considering the characteristics of condensate, keeping formation pressure higher than dew point pressure can minimize the loss of condensate oil; 5) considering the mechanism of gas injection displacement, slug gas injection can form pressure pulse, expand the sweep volume and improve the recovery of condensate oil. Those findings can also be used to guide the development of other gas condensate reservoirs.

Key words: buried-hill fracture; condensate gas reservoir; stress sensitive characterization; reservoir numerical simulation; cyclic gas injection development

● 0 Introduction

BZ condensate gas field is the largest natural gas field in Bohai Sea, with proven natural gas reserves of over 100 billion cubic meters and proven condensate reserves of more than 100 million cubic meters. The gas field has the following geological characteristics: buried deep(-4400~-5500m), complex seepage (fracture development), stress sensitivity (permeability loss 89.2%), and the fluid characteristics are as follows: low pressure difference between formation pressure and dew point pressure (1.07MPa), ultra-high condensate (711g / m3), large amount of reverse condensate (32.3%). The efficient development of BZ condensate gas field is a world problem！ It is found that there is no similar gas field development case at home and abroad, so it is impossible to accurately evaluate the development effect of this gas field. Therefore, it is urgent to carry out reservoir numerical simulation research to provide support for early development plan and subsequent efficient development[1-3].

● 1 Research method

According to the reservoir classification, the complex seepage system gas reservoir model is established. Based on the stress sensitive experiment, the variation law of seepage capacity is regressed. Based on the matching of fluid phase, the dual medium component model is established. And the development mode and gas injection development strategy are studied through reservoir numerical simulation.

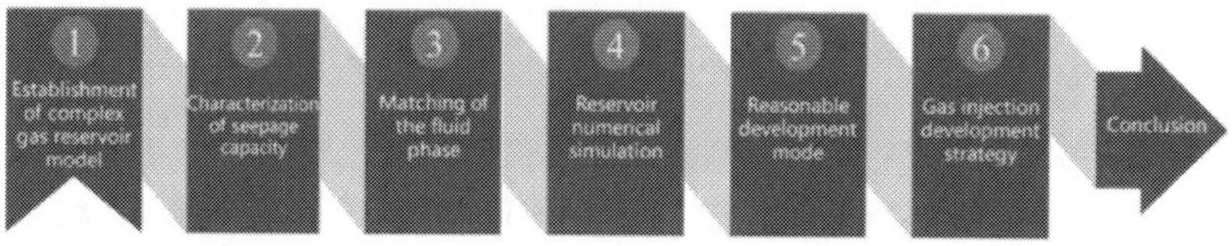

Fig 1. Technology Roadmap

● 2 Gas reservoir model

2.1 Reservoir zonation

Combined with the analysis of logging, core and testing data, BZ gas field has good reservoir connectivity, and it can be divided into sandy conglomerate zone, weathered zone and buried hill inner zone from top to bottom[4].

Fig 2. Vertical development characteristics of BZ condensate gas field reservoir

2.2 Establishment of complex gas reservoir model

Establish a structural model based on the understanding of reservoir zoning; and using lithofacies modeling and discrete fracture network modeling to generate matrix and fracture porosity, permeability and interporosity flow coefficient. By zeroing the matrix conductivity inside the buried hill, the coexistence of three types of complex seepage models in the BZ gas field can be realized.

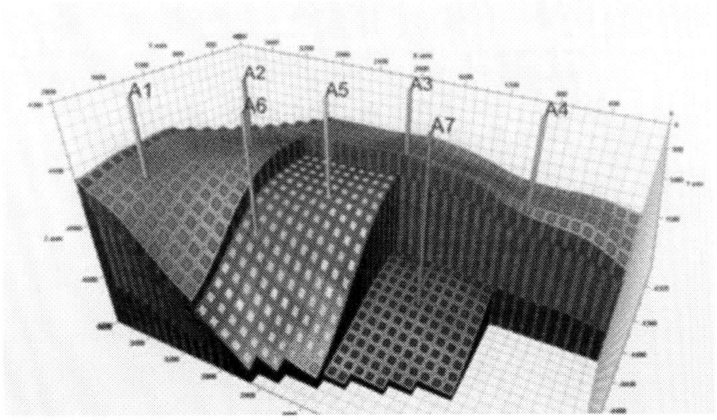

Fig 3. Gas reservoir model of complex seepage system

3 Stress sensitive characterization

3.1 Stress sensitivity analysis

According to the strong stress sensitivity of BZ gas field, the relationship between core permeability residual coefficient and net stress is fitted, and there is a good power function relationship between them; with the increase of net stress (reservoir pressure decreases), the rate of permeability loss gradually decreases[5-11]. In early stage, the net stress increased to 10MPa, Kr reduced by 60%; in mid stage, the net stress increased to 20MPa, Kr reduced by 20%; when reach the abandonment pressure(the net stress increased to 30MPa), the remaining permeability is about 10%.

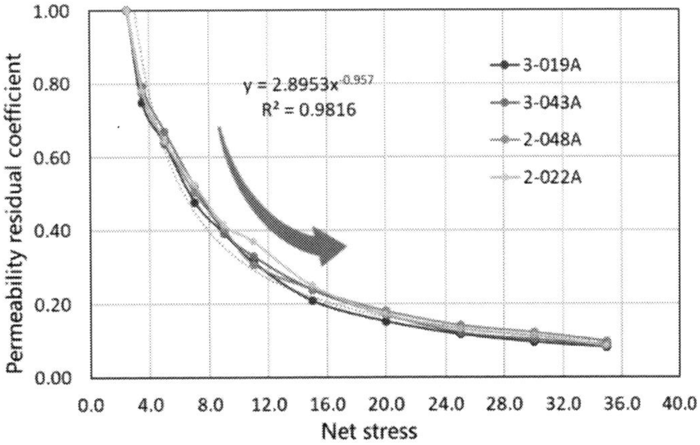

Fig 4. Permeability versus stress curve

$$K_r = 2.8953\ (P - P_0)^{-0.957}$$

K_r: Ratio of permeability to initial permeability, zero dimension;

P: Reservoir pressure at any time, MPa;

P_0: Initial formation pressure, MPa。

3.2 Stress sensitive characterization

According to the regression relationship, the characterization code of reservoir physical properties with time is compiled, and the sensitivity analysis is carried out by reservoir numerical simulation.

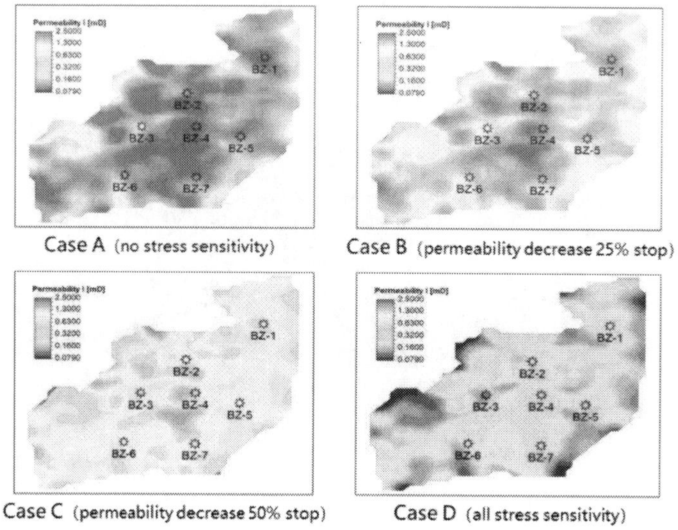

Fig 5. Comparison of permeability of different stress sensitive cases

● 4 Characterization of fluid characteristics

4.1 Fluid phase analysis

The BZ gas field is a condensate gas reservoir with high saturation and ultra-high condensate content through the study of fluid phase behavior characteristics, and the maximum condensate oil saturation is as high as 41%. When the pressure drops about 15MPa, the maximum saturation can be reached. Condensate oil precipitation seriously pollutes the reservoir, increases gas phase seepage resistance, and reduces gas well productivity and recovery[12-13].

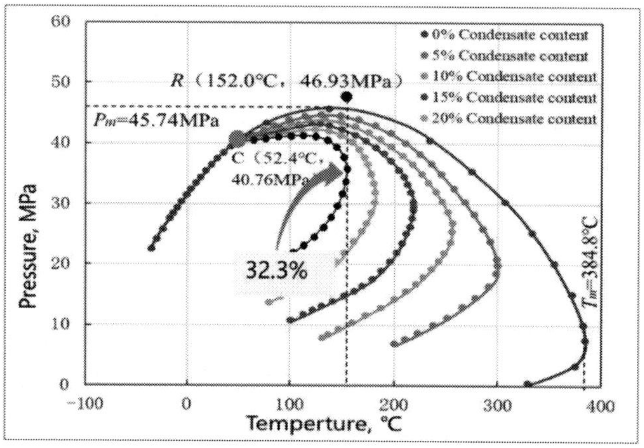

Fig 6. Phase diagram of fluid temperature and pressure in BZ gas field

4.2 Fluid phase matching

Considering the accuracy and speed of numerical simulation, 4 cases of component splitting combinations were set according to the principle of similarity combination. Combined with the calculation time and results of Numerical simulation of component model, the 7-component splitting plan is determined as the best plan.

11 components: C1、N2、C2、CO2、C3、C4、C5、 C6、C7-C10、C15+C22、C23+

7 components: C1+N2、C2+CO2、C3-C4、C5-C6、C7-C10、C15+C22、C23+

5 components: N2、CO2、C1、C2- C6、C7+

3 components: C1+N2、C2-C6+CO2、C7+

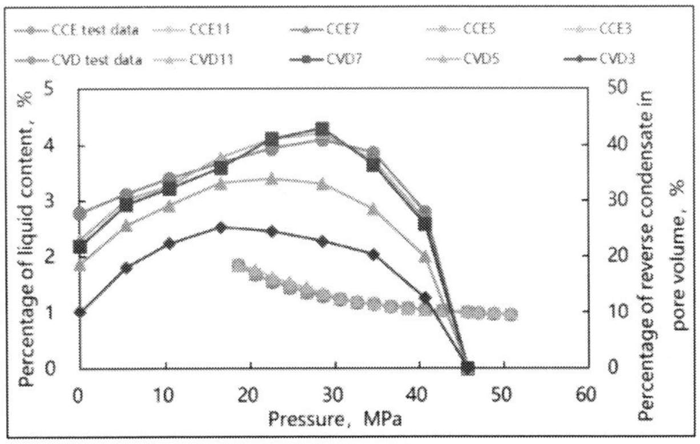

Fig 7. Comparison of experimental fitting accuracy

● 5 Study on gas reservoir numerical simulation

5.1 Optimization of development mode

As oil and gas fluids are produced, the formation pressure decreases, the physical properties become worse, the seepage ability decreases, and the recovery of gas field decreases. In depletion exploitation mode, the formation pressure drops rapidly, the condensate oil quickly reaches the maximum condensate velocity, and lost in the formation, so the condensate oil recovery rate is low. In gas injection development mode, the formation pressure is maintained through gas injection, and the retrograde condensate rate is controlled, and the condensate oil recovery is high. Considering reservoir stress sensitivity and retrograde condensate characteristics, from the perspective of improving oil and gas recovery and reservoir protection, it is recommended that the BZ condensate gas field development adopts first cycle gas injection and then depletion development[14-17].

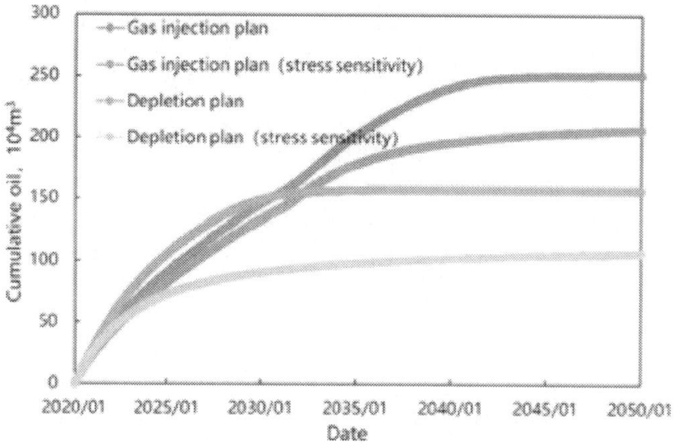

Fig 8. Comparison of different development mode

5.2 Cycle gas injection development strategy

Condensate gas reservoirs are usually developed with gas injection. The main purpose is to supplement formation energy, reduce condensate oil precipitation, and improve condensate oil recovery. When evaluating the effect of development, gas injection sweep volume and formation pressure maintenance level are two very important factors.

5.2.1 Optimization of injection and production

During gas injection development, the gas injection well position is directly related to the dry gas sweep volume. The structural and the fracture trend are two important factors.

A. Structural factor

The sandy conglomerate zone in structural high area of the BZ gas field has good physical properties. Priority is given to production in the early stage, and gas injection at the low area. In later stage, the condensate oil is enriched in the structural low area by gravity, and the top gas injection forms secondary gas and top drive. Vertical displacement, effectively expanding the gas sweep volume.

Tab 1. Development effect of structural factor

Plan	Cumulative gas 10^8m^3	Gas recovery %	Cumulative oil 10^4m^3	Oil recovery %
Inject in low area 10 years	36	53	181.3	26.6
Inject in low area 5 years,And in high area 5 years	36.6	53.3	192.4	28.3
Inject in high area 10 years	36.1	53.1	189.2	27.8

B. Fracture trend factor

When the injection-production line is parallel to the fracture trend, it is easy to form gas channeling, which reduce oil and gas recovery. When the injection-production line is vertical to the fracture trend, it can maximize the sweep volume, and effectively maintain formation energy and the seepage ability, and inhibit condensate oil to separate out.

Tab 2. Development effect of fracture trend factor

Plan	Cumulative gas 10^8m^3	Gas recovery %	Cumulative oil 10^4m^3	Oil recovery %
The injection-production line isparallel to the fracture	35.1	51.1	168.6	25.1
The injection-production line isperpendicular to the fracture	36.6	53.3	192.4	28.3
The injection-production line isoblique to the fracture	36	53	186.1	27.3

5.2.2 Reasonable pressure maintenance level

The most important thing in the development of condensate gas reservoirs is control the formation pressure to slow down the precipitation of condensate oil. The timing of gas injection and the rate of reinjection are two key factors.

A. Timing of gas injection

The timing of gas injection determines the degree of condensate oil loss, seepage capacity, and condensate precipitation rate during gas injection development. Since the difference between the formation pressure and the dew point pressure of the BZ gas field is very small, gas injection at the beginning of production can effectively delay the retrograde condensation time and reduce pollution, which is the optimal solution.

Tab 3. Development effect of timing of gas injection

Plan	Cumulative gas 10^8m^3	Gas recovery %	Cumulative oil 10^4m^3	Oil recovery %
Formation pressure 30MPa (4 years later)	26.4	48	127	23.4
Formation pressure 40MPa (2 years later)	26.3	47.9	132	24.4
Formation pressure 50MPa (0years later)	26.3	47.9	135.9	25.1

B. Re-injection rate

The higher the gas re-injection rate, the more stable the formation pressure remains. Before the formation pressure is higher than the dew point pressure, the 100% re-injection rate can always maintain the formation pressure higher than the dew point pressure, minimize the loss of condensate oil, and increase condensate oil recovery rate.

Tab 4. Development effect of Re-injection rate

Plan	Cumulative gas 10^8m^3	Gas recovery %	Cumulative oil 10^4m^3	Oil recovery %
100% re-injection rate	36.2	52.7	198.5	29.2
80% re-injection rate	36.4	53	196.1	28.8
50% re-injection rate	36.6	53.3	192.4	28.3

5.2.3 Optimization of gas injection mode

Considering the mechanism of gas injection and the demand for gas supply in the Beijing-Tianjin-Hebei region in winter, a comparative study of 3 plans: cyclic gas injection, slug gas injection, and Inject gas and soak. Circulating gas injection is prone to gas channeling, thereby reducing the condensate oil and gas recovery rate. The plan of inject gas and soak delays the gas channeling time on the basis of maintaining formation pressure, so the recovery rate is higher. In slug gas injection plan, the pressure pulse formed by the gas injection and gas production cycle can effectively expand the swept volume. And when injection stopped, the condensate in matrix will flow into the fracture due to the gravity. Therefore, the slug gas injection plan has the highest oil and gas recovery.

Tab 5. Development effect of Optimization of gas injection mode

Plan	Cumulative gas 10^8m^3	Gas recovery %	Cumulative oil 10^4m^3	Oil recovery %
Cyclic gas injection	35.3	50.2	188.4	26.8
Slug gas injection	37	53.9	198.4	29.1
Inject gas and soak	36.5	52.3	194.7	28.2

6 Conclusion

（1）The BZ gas field are divided into sandy conglomerate zone, weathered zone and Buried hill inner zone. The gas reservoir model of complex seepage system is established by the zeroing matrix conductivity inside the buried hill.

（2）The BZ gas field has strong stress sensitivity, and Nearly 90% loss of seepage capacity under abandonment pressure. Therefore, it is necessary to carry out numerical simulation research through time-varying reservoir physical properties.

（3）BZ gas field is a gas field with high saturation and high condensate content. Through PVTi matching and numerical simulation sensitivity analysis, the 7-component splitting plan is determined as the best plan.

（4） The numerical simulation study of the BZ gas field shows that: firstly inject gas in low area of the structure and then in the high area, inject gas at the start of production, the injection rate is 1.0, and seasonal slug gas injection is the optimal plan. This research provides strong support for

ODP program preparation and subsequent efficient development.

Reference:

[1] Fang Na, Jiang Guanghong, Li Yunpeng, et al. Development characteristics and potential tapping direction of buried-hill fractured oil reservoirs[J]. Special Oil & Gas Reservoirs, 2017,24(3):90-94.

[2] Xu Chunqiang, Zhang Zhen, Zhang Xintao, et al. Tectonic evolution and hydrocarbon accumulation model of western sag in western Bohai sea[J]. Journal of Northeast Petroleum University, 2018, 42（1）: 68-76.

[3] Zhang Feifei. Characterization and evaluation of pore-throat structures in ultra-low permeability and tight sandstones and their impacts on petrophysical properties: a case study of the lower Cretaceous in northern Songliao Basin[J]. Journal of Northeast Petroleum University, 2018,42(5):54-62.

[4] Tong Kaijun, Zhao Chunming, Lv Zuobin, et al. Petroleum exploration and development[J]. Petroleum Exploration and Development, 2012,39(1):56-63.

[5] Lin Guangrong, Liu Xuegang, Lu Yan, et al. Impact of stress change on percolation pattern of two-phase in low permeability reservoir[J]. Special Oil & Gas Reservoirs, 2008,15(1):43-45.

[6] Yang Bin, Hou Dali, You Gang, et al. Research and application of stress sensitive experiment on low permeable reservior[J]. Reservoir Evaluation and Development, 2012,5(2):41-45.

[7] Cao nai, Lei Gang. Stress sensitivity of tight reservoirs during pressure loading and unloading process[J]. Petroleum Exploration and Development, 2019,46(1):132-138.

[8] Jiang Ruizhong, Huang Lei, Cui Yongzheng, et al. Transit pressure analysis of a horizontal well in fractured composite reservoirs with stress-sensitivity effect[J]. Journal of Northeast Petroleum University,2018,42（3）: 84-91.

[9] Fang Yang. Research on relationship between deliverability of gas reservoir and the stress-sensitiveness of the reservoir[J]. Well Testing,2009,18（6）:18-21.

[10] Fu Lanqing. The effect of fractures on stress sensitivity and seepage characteristics in sandstone reservoirs[J]. Journal of Yangtze University（Natural Science Edition）,2016,13（20）: 14-18.

[11] Zhang Rui, Ning Zhengfu, Zhang Haishan, et al. New insights and discussions on stress sensitivity of fractured tight reservoir[J]. Natural Gas Geoscience, 2016,27(5):918-923.

[12] Wang Lezhi, Qi Zhilin, Jiang Yiwei. Study on retrograde condensate characteristics of deep condensate gas reservoirs with low permeability [J]. Natural Gas Technology, 2007, 1（1）: 52-54.

[13] Xiao Lixian, Du Jianfen, Guo Ping, et al. Pollution caused by retrograde condensate in condensate gas reservoir[J]. Fault-block Oil & Gas Field, 2009, 16（4）: 102-104.

[14] Cheng Liang, Zhang Xiaodong, Wang Juan, et al. Study on periodic gas injection mechanism of condensate gas reservoirs with high to high-super content of condensate oil in central Tarim by numerical simulation[J]. Journal of Xi'an shiyou University (Natural Science Edition),2016,31(2): 57-64.

[15] Wang Jing, Liu Huiqing, Liu Renjing, et al. Numerical simulation for low-permeability and extra-low permeability reservoirs with considering starting pressure and stress sensitivity effects[J]. Chinese Journal of Rock Mechanics and Engineering, 2013,32(2):3317-3327.

[16] Li Yong, Zhang Jing, Wu Xueyong, et al. A new reservoir simulation approach for fractured gas-condensate reservoirs[J]. Petroleum Exploration and Development, 2010,37(5):592-595.

[17] Jiang Hai, Yang Zhaozhong, Hu Yuehua, et al. Optimization of the cyclic gas injection parameters in condensate gas reservoir[J]. Journal of Xi'an shiyou University (Natural Science Edition),2009,24（2）:54-56.

AUTHOR INDEX